Sebastian Dreier
The Impact of Films on the Long-Term Behavior of Stationary Electrical Connections and Contacts in Electric Power Systems

TUDpress

Sebastian Dreier

The Impact of Films on the Long-Term Behavior of Stationary Electrical Connections and Contacts in Electric Power Systems

TUD*press*
2016

Die vorliegende Arbeit wurde am 05. Oktober 2015 an der Fakultät Elektrotechnik und Informationstechnik der Technischen Universität Dresden als Dissertation eingereicht und am 18. Dezember 2015 verteidigt.

Vorsitzender:
Prof. Dr.-Ing. Steffen Bernet

Gutachter:
Prof. Dr.-Ing. Steffen Großmann
Prof. Dr. Noureddine Ben Jemaa
Dr.-Ing. habil. Konstantin Papailiou

Bibliografische Information der Deutschen Nationalbibliothek
Die Deutsche Nationalbibliothek verzeichnet diese Publikation in der Deutschen Nationalbibliografie; detaillierte bibliografische Daten sind im Internet über http://dnb.d-nb.de abrufbar.

Bibliographic information published by the Deutsche Nationalbibliothek
The Deutsche Nationalbibliothek lists this publication in the Deutsche Nationalbibliografie; detailed bibliographic data are available in the Internet at http://dnb.d-nb.de.

ISBN 978-3-95908-037-8

Eckhard Richter & Co. OHG
Bergstr. 70 | D-01069 Dresden
Tel.: 0351/47 96 97 20 | Fax: 0351/47 96 08 19
http://www.tudpress.de

TUDpress ist ein Imprint von w. e. b.

Gesetzt vom Autor.
Printed in Germany.

Technische Universität Dresden

The Impact of Films on the Long-Term Behavior of Stationary Electrical Connections and Contacts in Electric Power Systems

Sebastian Dreier

von der Fakultät Elektrotechnik und Informationstechnik

der Technischen Universität Dresden

zur Erlangung des akademischen Grades eines

Doktoringenieurs

(Dr.-Ing.)

genehmigte Dissertation

Vorsitzender:	Prof. Dr.-Ing. Steffen Bernet		
Gutachter:	Prof. Dr.-Ing. Steffen Großmann		
	Prof. Dr. Noureddine Ben Jemaa	Tag der Einreichung:	05.10.2015
	Dr.-Ing. habil. Konstantin Papailiou	Tag der Verteidigung:	18.12.2015

Acknowledgement

This research work is the result of my research activities at the Institute of Electrical Power Systems and High Voltage Engineering of the Technische Universität Dresden.

First of all, I want to thank my advisor Prof. Dr.-Ing. Steffen Großmann for encouraging me to start this research project, the good work environment and scope for development at our institute as well as his constant willingness to discuss any problems that have occurred along the way.

I would also like to thank Prof. Dr. Noureddine Ben Jemaa for the acceptance of the thesis report as well as for offering me the opportunity to stay at the Institut de Physique de Rennes in 2012. I am very grateful for both and also for the support of Elsa, Hans and Erwann during my time in Rennes.

In addition to that, I wish to thank Dr.-Ing. habil. Konstantin Papailiou for serving on the doctoral committee and for all of his contributions made as well as for the inspiring talks.

Moreover, I want to thank Dr.-Ing. Nils Lücke for his mentorship which provided many valuable hints and advice for my work. I would also like to thank Dr.-Ing. Stephan Schlegel for his continuous support, his guidance and several inspiring impulses during the doctoral process.

I also want to thank Benjamin Kaufmann for the great time we spent together at the institute including our collaborative projects and research activities, Michael Gatzsche for the very inspiring office community and Christian Hildmann for the helpful talks.

I would like to thank all other staff members of our institute that were somehow engaged in this work I have not mentioned yet. Moreover, I have to thank all my former students who I have supervised as they have all somehow supported this project in their own way.

Furthermore, I want to thank all project partners in Dresden (Botanical Garden), Egypt (Dr. Ghareeb Moustafa from Suez Canal University), Brazil (Diego Guimarães dos Santos, Prof. Macedo, Prof. Rocha from Sistema de Proteção da Amazônia) and any other persons or institutions that I have not mentioned by name so far.

Lastly, I would like to thank my family and especially my sister Tina for her valuable linguistic support. And finally, I want to thank Alexandra for her continuous support and love, always standing by my side and enriching my life.

Dresden, February 2016

Sebastian Dreier

Abstract

Stationary electrical connections and contacts, such as power connections, are commonly applied in electric power systems used for generation, transmission and distribution of electric energy. Several different degradation mechanisms can increase the contact resistance and might therefore reduce the power connection's lifetime. The degradation by film development as a result of chemical reactions is often considered as a reason for contact failure.

In this research work, the impact of film development produced by chemical reactions, such as oxidation, on the long-term behavior of stationary electrical connections and contacts was studied with crossed rods. Analytical, numerical and experimental methods were applied. Typical material systems for electric power systems were considered in this study: Cu-ETP (CW004A) bare, silver-, nickel- or tin-coated, Al99.5 (EN AW-1050A) and AlMgSi0.5 (EN AW-6060).

By applying numerical methods, the mechanical stress distribution was determined within a circular contact point. The initial contact resistance and the plastic deformed area of the considered material systems was measured in experimental tests. The film's impact was further determined through comparative experimental studies in air (standard atmosphere) and N_2 (inert gas).

During the experimental tests on perpendicularly crossed rods, other degradation mechanisms such as force reduction were suppressed. The film's impact within the formation phase was studied on copper rods in an oven at 200 °C for 1000 h. Moreover, the dependency on different environments at 90 °C (laboratory, botanical garden and outdoor) was tested for 12000 h. Additional long-term tests over 12000 h were conducted at 200 °C. The contact resistance was determined dependent on time. Furthermore, the plastic deformed area was ascertained by microscopy. It was found that the time dependent film development caused by chemical reactions such as oxidation might possibly not result in a significant degradation of stationary electrical contacts with circular contact points and a constant force.

Supplementary studies were performed at 200 °C for 1000 h with perpendicularly crossed rods at low forces (3.5 N) as well as analytical assessment of radial and axial film growth on circular contact points. The measured long-term behavior of perpendicularly crossed rods was similar for low and high forces.

In order to study the long-term behavior of power connections operated in areas with harsh environmental conditions, experimental field tests on bolted busbar joints were conducted in desert and tropical rainforest environments. For over two and a half years, long-term field tests investigating bolted busbar joints made of Cu-ETP, Al99.5 (EN-AW-1350A) or AlMgSi0.5 (EN AW-6060) either with or without coating (silver, tin or nickel) were conducted in Belém (Brazil), Ismailia (Egypt) and Dresden (Germany).

Table of Contents

List of Abbreviations and Symbols

Symbol	Definition	Unit
a	Spot radius	µm
α	Semi axis ellipsoid	-
α_c	Cluster radius	µm
α_T	Temperature coefficient electric resistance	1/K
α_{th}	Coefficient of linear thermal expansion	1/K
A	Area	mm^2
A_a	Apparent contact area	mm^2
A_b	Load-bearing contact area	mm^2
A_w	Current-carrying area	mm^2
A_f	Film-covered area	mm^2
A_i	Zonal area *i*	-
A_m	Metallic contact area	mm^2
A_q	Quasi-metallic contact area	mm^2
β	Semi axis ellipsoid	-
BR	Brazil	-
c	Concentration	-
d	Diameter	mm
D	Diffusion coefficient	m^2/s
D_0	Frequency factor	m^2/s
DE	Germany	-
e	Elementary charge	C
E	Elastic modulus	GPa
E_t	Tangent modulus	kN/mm^2
ε_r	Relative permittivity	-
ε	Strain	-
EG	Egypt	-
F	Force	N
Φ	Work function	eV
$\varphi_{\mu,\sigma}(x)$	Densitiy function	-
h	Planck constant	Js
H	Hardness	N/mm^2
H_f	Film hardness	-
H_K	Contact hardness	GPa
H^*	Bulk material hardness	-
I	Current	A
I_i	Current in zonal area *i*	-
I_{ia}	Current per spot in zonal area *i*	-

J	Current density	A/mm^2
J_i	Current density in zonal area i	-
J_{ia}	Current density per spot in zonal area i	-
k_B	Boltzmann constant	$kg \cdot m^2 \cdot s^{-2} \cdot K^{-1}$
k_u	Performance factor	-
$k_{u,0}$	Initial performance factor	-
$k_{u,crit}$	Critical performance factor	-
K	Sphere K	-
κ	Conductivity	$S \cdot m^{-1}$
l	Half distance between a-spot center	µm
l_{busbar}	Length busbar sequence	mm
l_J	Overlapping length busbar joint	mm
λ	Thermal conductivity	W/(m·K)
l	Length	mm
m_e	Electron mass	kg
$\overline{\mu}$, μ_φ	Arithmetic mean	-
$\sqrt{\mu}$	Height semi ellipsoid equipotential area	-
n	Number of a-spots / microcontacts	-
n_φ	Number of measurements	-
v	Poisson's ratio	-
m	Exponent deformation	-
ξ	Shape factor	-
p	Average pressure	Pa
p_0	Maximum pressure	Pa
P_J	Power losses busbar joint	W
P_{busbar}	Power losses busbar sequence	W
Q_a	Activation energy	kJ/mol
r	Radius	µm
r_t	Radius plastic deformed area	µm
r_{xy}	Radius x-y-area	-
r_{xyi}	Radius x-y-area in zone i	-
ρ	Resistivity	µΩ·mm
ρ_b	Resistivity bulk material	µΩ·mm
ρ_{coa}	Resistivity coating	µΩ·mm
ρ_f	Resistivity film	µΩ·mm
ρ_{20}	Resistivity at 20 °C	µΩ·mm
ϱ_{20}	Density at 20 °C	g/cm^3
R	Contact resistance	µΩ
R_{bulk}	Share bulk resistance busbar joint	µΩ

R_{busbar}	Resistance busbar sequence	µΩ
R_c	Constriction resistance	µΩ
R_f	Film resistance	µΩ
R_J	Connection resistance busbar joint	µΩ
$R_{p0.2}$	Yield point	N/mm^2
R_m	Ultimate tensile strength	N/mm^2
R'	Gas constant	$J \cdot K^{-1} \cdot mol^{-1}$
$R(T)$	Resistance measured at temperature T	µΩ
R_{20}	Resistance at 20°C	µΩ
RH	Relative Humidity	%
s	Thickness	-
s_f	Thickness coating	µm
s_f	Film thickness	nm
$s\uparrow$	Upper bound minus arithmetic mean	-
$s\downarrow$	Lower bound minus arithmetic mean	-
σ	Stress	-
σ_e	Equivalent stress according to von Mises	MPa
σ_f	Tunnel film resistivity	$\mu\Omega \cdot mm^2$
σ_φ	Standard deviation	-
σ_{ij}	Individual stress component	-
t	Time	h
T	Temperature	K
T_{20}	Temperature at 20 °C	K
ΔT_{amb}	Ambient temperature spread	K
ϑ	Temperature	°C
ϑ_{amb}	Ambient temperature	°C
ϑ_m	Melting point	°C
ϑ_{max}	Max. operating temperature	°C
ϑ_s	Softening point	°C
U_{AB}	Voltage drop across cylinder AB	V
U_{busbar}	Voltage drop busbar sequence	V
U_J	Voltage drop busbar joint	V
$\parallel$	Parallelism	-

1 Introduction

The global energy generation currently faces a large transition from fossil to renewable energy sources. The efficiency of the energy generation through renewables is particularly depending on their location. For instance, permanently good wind conditions for windmills can be found on sea whereas excellent radiation levels for solar thermal or photovoltaic power plants are mostly found in dry and hot deserts. The proposed construction sites of many renewable energy generation facilities are situated as close as possible to major consumption areas in order to be implemented into already existing grid solutions. In many countries the most promising sites for the generation of renewable energy are far away from demanding consumption centers. In order to connect load and generation areas, electric energy has to be transmitted and distributed over long distances, often through areas with harsh environmental conditions. Furthermore, large upcoming energy projects, such as hydropower plants or the electrification of large areas in emerging countries, focus on regions lacking the appropriate energy infrastructure. Consequently, all electrical applications used in such regions not only have to withstand extreme conditions, but they also have to maintain good performances in order to assure the reliable and efficient supply of energy.

Stationary electrical connections and contacts applied in electric power systems/power engineering have to fulfill high reliability requirements to prevent black outs in the grid or in other components of the power system. In terms of operation, very harsh environmental conditions affecting stationary electrical connections and contacts can be observed. These conditions are typically characterized by high values of environmental parameters. This study aims to investigate the performance of technical stationary connections and contacts applied in electric power systems with respect to environmental conditions. Therefore, long-term field tests of stationary power connections in areas with harsh environment conditions were conducted.

The aging of stationary electrical connections and contacts is caused by a number of different degradation mechanisms, namely interdiffusion, force reduction, electromigration, sliding wear and fretting as well as chemical reactions (Fig. 1).

Through the aging processes the contact resistance rises and eventually results in higher Joule heating as well as higher contact temperatures. This process continues until the contact resistance and the contact temperature are too high (thermal runaway) and the contact fails. Individual degradation mechanisms, such as film development caused by chemical reactions, are often considered as the main reason for contact or connection failure in electric power systems.

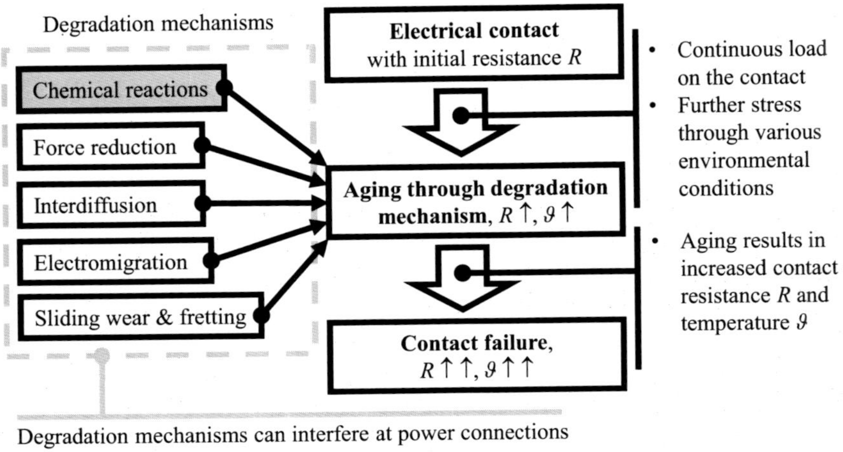

Fig. 1: Overview of degradation mechanisms and interaction with electrical contacts

Unfortunately, other degradation mechanisms cannot be excluded for the majority of the known investigations and technical applications. Consequently, the question on whether individual degradation mechanisms, such as film development, influence the aging of stationary electrical connections and contacts arises. The purpose of this research study is to investigate the degradation mechanism due to film development by chemical reactions, such as oxidation, on stationary electrical connections and contacts applied in electric power systems. Typical materials applied in power engineering are copper and aluminum and its alloys partially being improved by coatings. The considered materials in this study are copper (Cu-ETP), pure aluminum (Al99.5) and an aluminum alloy (AlMgSi0.5) which are typically used for devices in power engineering, overhead power lines or busbar systems. The materials were either bare or coated with silver, tin or nickel. Cleaning the interface area before being assembled is a standard procedure for power connections in electric power systems and was applied in this study. Within this research work, experimental tests were conducted with reference to the long-term behavior of electrical connections and contacts as well as the degradation mechanism chemical reactions. Besides the experimental work, microscopic studies investigating the deformed contact area and material investigations conducted at different stages and conditions of the considered contacts were further applied.

2 Basics of Electrical Contacts

2.1 Theory and Principles

If two conductors, called contact members (conductor 1 and 2, Fig. 2), contact each other at a mutual interface which is able to carry electric current, an electrical contact is established. Usually the contact members are pressed together by a load or force F. Depending on the shape, dimensions, material properties of the contact members and the force F the contact area is established. Due to manufacturing processes the electrical connections in power engineering are characterized as form bonded, force-fit or firmly bonded. The whole interface area of conductor 1 and 2 is termed apparent contact area A_a and consists of different subareas.

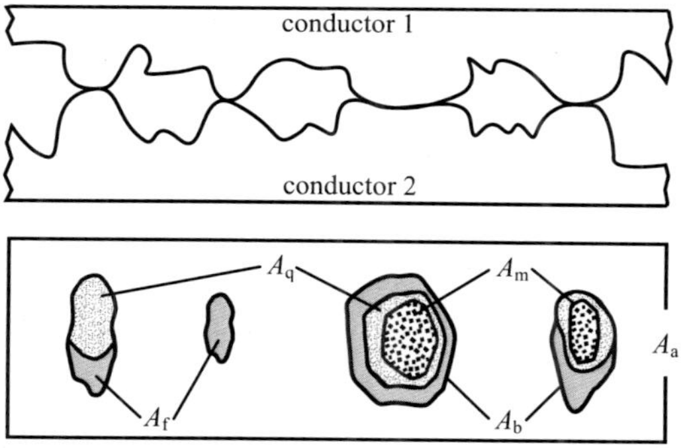

Fig. 2: Sketch of contact area

Because of the surface roughness of technical surfaces the contact members only touch each other at a number of spots that form the load-bearing area A_b. An individual spot is termed a-spot and can be treated as individual electrical contact while a larger number of a-spots are forming a multiple contact. Within the load-bearing area A_b a metallic area, a quasi-metallic and an insulation area exist. The metallic contact area A_m is formed by pure metallic contacts whereas the quasi-metallic area A_q is covered by a pollution film passable by electrons. If the film becomes thicker it then insulates the interfaces of the contact members. Depending on the film's properties these film-covered area A_f can be semiconducting up to insulating. Usually the electric current can flow only through the a-spots across the metallic and the quasi-metallic areas [1]. Therefore, the current-carrying area A_w is the metallic contact area A_m and in certain case it is supplemented by the quasi-metallic area A_q. The lower the current-carrying area A_w the higher the connection or contact resistance.

Before the current passes the contact area the current flow lines will be constricted in the bulk material of the conductors. Thereby the current lines pass equipotential surfaces. The mode of constriction in homogenous material with homogenous current feeding depends on the geometric characteristic of the contact members. In conductors with non-homogenous materials the conductivity also affects the current spreading.

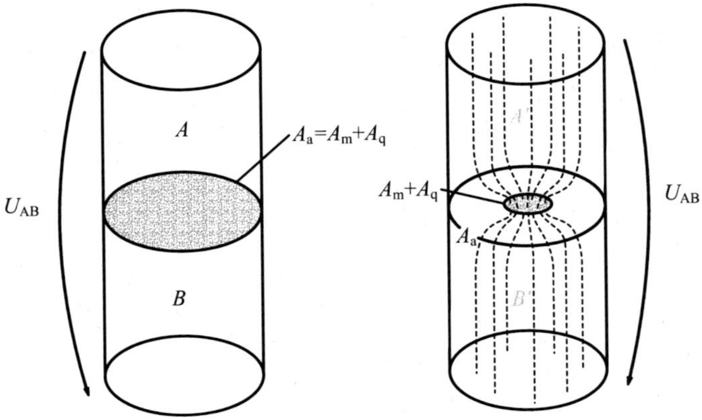

Fig. 3: Cylinder *AB* with no constriction and cylinder *A'B'* with constricted current flow (dashed lines) through conductive areas

As shown above (Fig. 3), cylinder *AB* is an ideal conductor with no constriction with a perfect conducting apparent contact area. The current flow for cylinder *A'B'* is constricted to an area smaller than the cross section of the cylinder. Holm [1] defines the electrical contact resistance as the resistance caused by a voltage drop between the equipotential areas of the contact members and the current flow (cylinder *A'B'*) minus the resistance of a conductor sequence with the same length and no constriction (cylinder *AB*). Therefore, $R_{AB} = U_{AB} / I$ and constriction resistance $R_c = R_{A'B'} - R_{AB}$. Holm [1] deploys two mathematical methods to describe the constriction resistance of an individual a-spot : the spherical and the ellipsoid model.

Sphere Model

The spherical model assumes a sphere K with a radius a and an infinite conductivity (Fig. 4). When assuming symmetrical and infinite conductors, the equipotential surfaces are hemispherical distributed across the conductor starting from the sphere K. The current I spreads uniformly with current density J. For both contact members a constriction occurs that can be summarized in a mutual constriction resistance R_c with radius a and resistivity ρ (Tab. 1).

Ellipsoid Model

The ellipsoid model assumes a plane ellipse with semi-axes α and β. Equipotential surfaces are described as semi-ellipsoid with the height $\sqrt{\mu}$ perpendicular to the contact area. If $\alpha = \beta$ a circular contact area results. When measuring the resistance of power connections, often significant bulk material shares are measured together with the contact resistance. By definition only the constricted volume of a conductor sequence belongs to the contact resistance and is proportional to μ. Although Holm defines the term: long constriction. When the constricted conductor sequence with no homogenous current distribution is essential larger than the volume a^3 around the a-spot and secondly the bulk resistance of an equivalent conductors sequence with homogenous current distribution can be neglected, a long constriction exists. The larger the constricted conductor sequence, the lower the impact of the bulk resistance for a conductor sequence with no constriction and the same taping points. An individual a-spot and with a long constriction can be defined as constriction resistance R_c (Tab. 1, (3)). For $\mu \rightarrow \infty$ and $\alpha = \beta$, (7) results (Tab. 1).

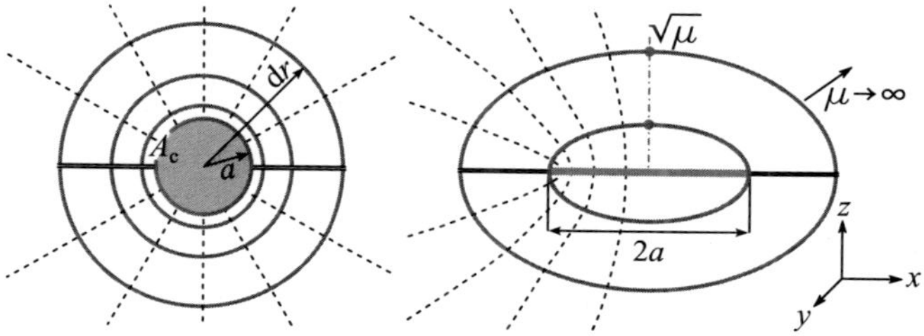

Fig. 4: Sphere model on the left with spot radius a, independent variable r and superconducting sphere & surface area A_c; ellipsoid model on the right with circular spot and radius a (view from z-direction) as well as semi ellipsoid equipotential areas with height $\sqrt{\mu}$

Tab. 1: Constriction resistance and current density for ellipsoid & sphere model [1]

Model	Constriction resistance R_c	Current density J
Sphere	$R_c = \frac{\rho}{\pi \cdot a}$ (1)	$J = \frac{I}{2\pi \cdot a^2}$ (2)
Ellipsoid	$R_c = \frac{\rho}{\pi \cdot a} \arctan \frac{\sqrt{\mu}}{a}$ (3)	$J(x,y) = \frac{I}{2\pi\alpha \cdot \beta} \cdot \frac{1}{\sqrt{1-\left(\frac{x^2}{\alpha^2}+\frac{y^2}{\beta^2}\right)}}$ (4)
Ellipsoid with $\alpha = \beta = a$	$R_c = \frac{\rho}{\pi \cdot a} \arctan \frac{\sqrt{\mu}}{a}$ (5)	$J(x,y) = \frac{I}{2\pi \cdot a} \cdot \frac{1}{\sqrt{a^2 - x^2 - y^2}}$ (6)
Ellipsoid with $\alpha = \beta = a$, long constriction $\mu \rightarrow \infty$	$R_c = \frac{\rho}{2a}$ (7)	$J(r) = \frac{I}{2\pi \cdot a} \cdot \frac{1}{\sqrt{a^2 - r^2}}$ (8)

In general, a thin film is normally present on the surface of metallic conductors. If the film in Fig. 5 between cylinder *A'* and *B'* is not penetrated, the film resistance R_f has to be added to the bulk resistance of the cylinder *AB*. For an individual a-spot, covered with a highly resistive film $\rho_f >> \rho$ and a given long constriction the current is uniformly distributed across the film as illustrated in cylinder *A'B'*. Beside the film resistance R_f the films could additionally influence the constriction resistance R_c. In cylinder *AB''* the current lines will seek low resistivity paths and eventually become constricted in areas with lower material resistance.

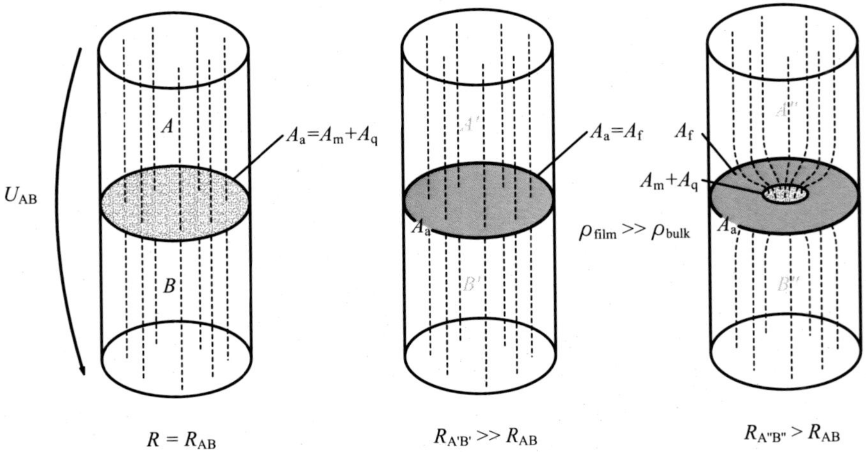

Fig. 5: Current distribution (dashed lines) across cylinder AB without film, cylinder AB' with entire film cover and cylinder $A''B''$ with partial film cover

If the film during the contact members cannot be penetrated and remains present a further resistance has to be added to the contact resistance. For the contact resistance of an individual a-spot the film resistance R_f is in series with the constriction resistance R_c.

$$R = R_c + R_f \tag{9}$$

The direct contact resistance measurement on technical stationary electrical connections, for instance bolted busbar joints, is often not possible. The measured connections resistance contains of a bolted busbar joint R_J contains some bulk shares and can be described as: $R_J = R_{bulk} + R_c + R_f$. The current conduction across the film takes place through two possible phenomena: the common transport mechanism or the tunnel effect. Very thin layers with a thickness of a few nm can be passed by charger carriers through quantum tunneling effect [1]. The literature is not clear about the maximum film thickness. When tunneling occurs, the tunnel resistance can be determined by the methods of Holm [1], Eqs. (12) – (14), Simmons [2] and Slade [3], Eq. (15), (Tab. 2). The determined film resistivity of Simmons approach [2] is between the approaches in [1], [3].

Tab. 2: Film resistance with respect to the conduction mechanism [1], [3] with film resistivity ρ_f, bulk resistivity ρ_b, film thickness s_f, tunnel film resistivity σ_f, Planck constant h, elementary charge e, electron mass m_e, work function Φ, relative permittivity ε_r

	Film resistance R_f if $\rho_f >> \rho_b$		
	Common transport mechanism	Tunnel effect	
[3]	$\frac{\rho_f \cdot 2s_f}{\pi \cdot a^2}$ (10)	$\frac{\sigma_f}{\pi \cdot a^2}$ (11)	[1]
	$s_f >> (2\ldots5)$ nm	$s_f \leq (2\ldots5)$ nm	
	Tunnel film resistivity σ_f		
	$\sigma_f = \left(10^{-22}/2\right)\cdot\left(A^2/(1+A\cdot B)\right)\exp\left(A\cdot B\right)$ (12)		
[1]	$A = 7.32\cdot10^6\left(s-0.72/\Phi\right)$ (13)		
	$B = 1.265\cdot10^{-6}\sqrt{\Phi-1/\left(s_f\cdot\varepsilon_r\right)}$ (14)		
[3]	$\sigma_f = \left(2s_f\cdot h^2/\left(3e^2\sqrt{2m_e\cdot\Phi}\right)\right)\cdot\exp\left(4s_f\cdot\pi\sqrt{2m_e\cdot\Phi}/h\right)$ (15)		

According to Holm [1] and Lawless [4] films can be distinguished dependent on composition, thickness and temperature. Pre-tarnish films consist of just an individual layer of adsorbed oxygen. Later on solid oxides develop to form tarnish films. Other films could be adsorbed by water, liquids or technical lubricants. At a clean metallic surface free oxygen components could be adsorbed due to physisoption induced by relatively weak forces defined as van der Waals forces. These bonds are reversible and require no activation energy. The binding energy is as low as 0.05 eV. This is nearly one to two orders below the binding energies between atoms in molecules [1]. Stronger covalent and mostly irreversible bonds between surface and oxygen develop through chemisorption. Depending of the self-limiting properties the films will either continue to grow or become passivating. Lawless classification rests upon temperatures and the temperature dependent film thickness [1]. Both classifications are merged in the figure below.

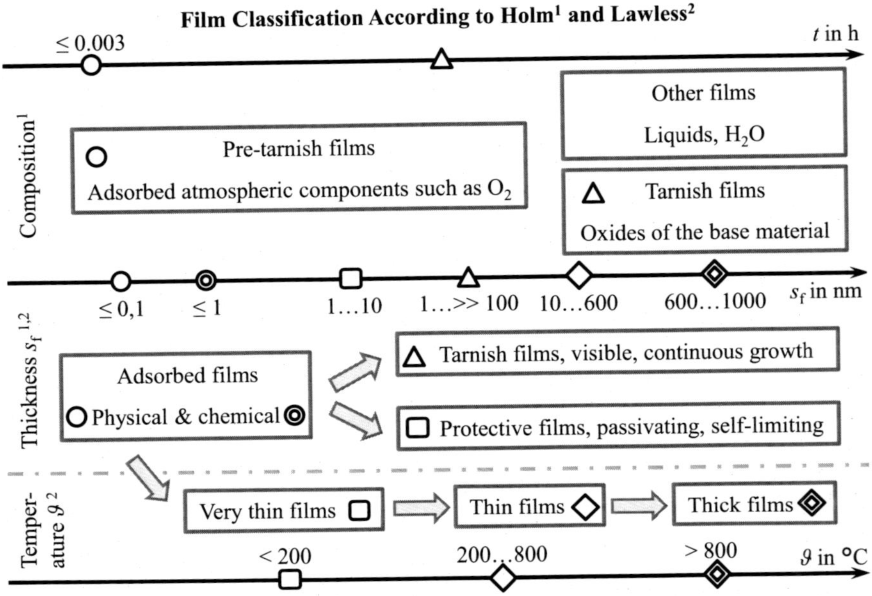

Fig. 6: Film classification (by symbols) dependent on film composition, thickness and temperature [1], [4]

It is known that initial films could increase the contact resistance for technical connections and contacts. High mechanical forces and proper surface cleaning will be applied for stationary connections and contacts in electric power systems. Thin films will be mainly penetrated or displaced by this actions. Nevertheless a qualitative and quantitative proof of this hypothesis still has to be developed.

Multiple Contact Spots

The surface of metallic materials used in electrical energy applications have a roughness even if they are polished and appear as smooth. Therefore, at the interface of the contact members a variety of contact points is established (Fig. 1). Depending on the location and the individual spot's properties the contact spots interact among themselves. The interaction disappears if the distance $2l$ lies between each spot and the ratio distance to spot radius l/a is large. In this case Holm assumes a long constriction and defines the constriction resistance as:

$$R_c = \frac{\rho}{2\sum a_i} \tag{16}$$

Here the contact spots interrelate like parallel-connected elements within an electrical network. The higher the number and the larger the a-spot radius, the lower the resulting resistance. The radius a is generally very small compared to the dimensions of a technical connection or contact. Therefore, the potential drop across the a-spots of technical connections and contacts is measured at a larger volume than the interface area itself. Thus the voltage measurement typically involves the constriction of the current in the conductors towards the contact spots where the current spreads across the individual a-spots. The constriction towards the contact area depends on the geometric characteristic of the contact members. For a circular contact area with radius α_c, n circular spots with equal radius a lay all within the cluster.

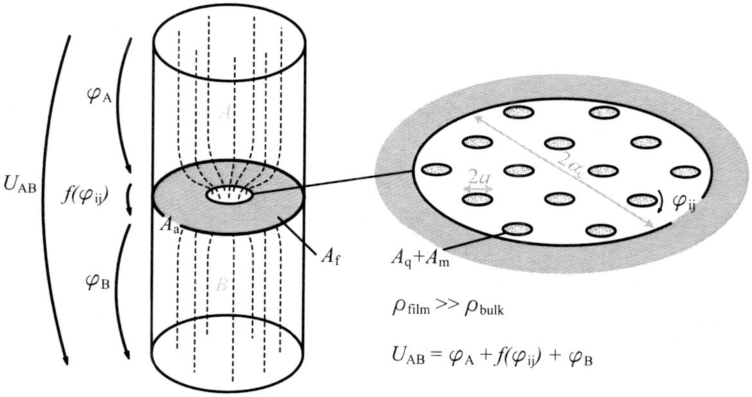

Fig. 7: Multiple contact spot interface between cylinder A and B and detail view of the contact interface

The problem can be described as capacitance problem of n charges of the a-spots if they all have a potential towards infinite in an infinite conductor. Every charge quantity per contact spot has its own potential φ_{ij} due to local constriction. Indeed all other contact spots within the cluster induce an additional potential in the individual spot. This potential can approximately be described as if all charges were distributed uniformly across the cluster [5]. For all contact spots with equal potential within the interface, the contact resistance can be defined as [5], [6]:

$$R = R_{ca} + R_{c\alpha} = \frac{\rho}{2n \cdot a} + \frac{\rho}{2\alpha_c} \tag{17}$$

Within the cluster the a-spots are evenly distributed. The cluster can be divided in further sub-clusters [7]. Holm indicates possible errors $\approx 10\,\%$ for the formula stated above if $a > \alpha_c / (3 \cdot \sqrt{n}) \wedge n < 3$.

For his model description (Fig. 7) Greenwood [6] assumes a cluster of n microcontacts with large distances to other clusters as well as large contact area. The potential at a contact is determined by the self-resistance of contact i and the potential between contact i and infinity induced by the current through contact j. The self-resistance can be influenced if the current redistributes in the presence of other contacts. For his model Greenwood assumes that microcontacts with distance s_{ij} between the contacts are far away from each other and act independently.

$$R = \frac{\rho}{2\sum a_i} + \frac{\rho}{\pi} \cdot \left(\sum\sum_{i \neq j} \frac{a_i \cdot a_j}{s_{ij}} \right) / \left(\sum a_i\right)^2 \tag{18}$$

The interaction is supplemented to the constriction resistance. For equal spot radii the equation simplifies to

$$R = \frac{\rho}{2n \cdot a} + \frac{\rho}{\pi \cdot n^2} \cdot \sum\sum_{i \neq j} \frac{1}{s_{ij}} \tag{19}$$

The introduced error becomes small for a large n. According to Greenwood the true contact area cannot be determined by measurement, except for an upper limit. He further states that the usage of the bulk hardness in order to determine the mechanical deformed area is not exact because of the non-consideration of the surface asperities.

2.2 Deformation and Plasticity

2.2.1 Elastic Deformation

When an external load is applied, stress appears which results in a deformation and a relative length extension. The linear relationship between stress and deformation can be described by Hooke's law:

$$\sigma_{\mathrm{ij}} = \varepsilon_{\mathrm{kl}} \cdot \sum_{\mathrm{kl}} c_{\mathrm{ijkl}} \tag{20}$$

Whereas c_{ijkl} are elastic constants, the components σ_{ij} and $\varepsilon_{\mathrm{ij}}$ belong to the stress and strain tensor. For an uniaxial stress state with isotropic material behavior the following holds true [8]:

$$\sigma = E \cdot \varepsilon \tag{21}$$

The elastic constant E is the elastic modulus and describes the slop of a curve in the stress-strain-plot until the yield point is reached at which plastic deformation occurs.

The elastic deformed contact area of two spheres or two cylinders with their axes at an right angle and a curved surface can be calculated with the approach of [9] stated in [1]:

$$r_t = \sqrt[3]{\frac{3}{4} F \cdot \left(\frac{1-v_1^2}{E_1} + \frac{1-v_2^2}{E_2} \right) \cdot \left(\frac{1}{r_1} + \frac{1}{r_2} \right)^{-1}} \tag{22}$$

In Eq. (22) the radius of the mechanical deformed area r_t is determined with the radii r_1 and r_2 of the corresponding bodies and the Poisson's ratios v and its elastic modulus E. According to Hertz [10] the pressure distribution $p(r)$ along the interface of a contact spot with radius a, variable r and a maximum pressure p_0 is determined by the following formula:

$$p(r) = p_0 \cdot \left(1 - \frac{r^2}{a^2} \right)^{1/2} \tag{23}$$

2.2.2 Plastic Deformation

If mechanical stress exceeds a certain level plastic deformation occurs. At this point the yield strength is exceeded and material displacement occurs. The material also undergoes a strain hardening process which could result in an increased local hardness. Plastic deformation is a non-reversible process and results in permanent deformations. At a crystalline scale the main reason for this process is the anisotropic slipping of atomic layers along other atomic levels due to significant high shear stress. The size and the slip distance of the atomic layers differ. Both can be described by the Burgers vector. The Burgers vector describes the direction of displacement but it is also an indicator for dislocations in the lattice. At dislocations, local defects and irregularity in the lattice occurs and are therefore susceptible for plastic deformation [8]. In reality the transition point between elastic and plastic deformation can hardly be determined. From a technical perspective, the yield strength is defined as elastic yield point $R_{p0.2}$ at which only 0.2 % of the strain ε remain after removing the applied load. The hardness H describes the resistance of a solid material against a test shape or indentor. The remaining indentation can be evaluated and derives the experimentally obtained hardness. Several methods exist to determine the hardness experimentally, e.g. Brinell, Vickers, Meyers. All depend on the applied force and the shape of the indentors. According to Holm the radius of the mechanical deformed area for plastic deformation can be expressed as follows:

$$r_t = \sqrt{\frac{F}{\pi \cdot H}} \tag{24}$$

Holm states, that because of the surface roughness both the indentations and also the level of deformation differ from each other. He introduces a factor ξ with ξ being between 0.1 and 0.3 and determines a lower plastic deformed area with $r_t = (F / (\pi \cdot \xi \cdot H))^{0.5}$. In the literature various equations exist to express the contact hardness H_K.

After the contact formation process a time dependent plastic deformation processes exist (Fig. 8). If the applied force remains high enough the material tends to undergo creep and stress relaxation (see also Fig. 9.). Dependent on temperature, stress and time, creep entails a change of strain into permanent plastic deformation. The macroscopic process can be divided in several stages. Nevertheless all of them result in additional strain rise and reduced spring range of the material. Stress relaxation is a material dependent process with the strain ε = const. The process is a permanent transformation of elastic strain into plastic strain. As a result, the mechanical stress will be reduced. Stress relaxation not only depends on time and temperature but it is more effective for higher temperatures [11].

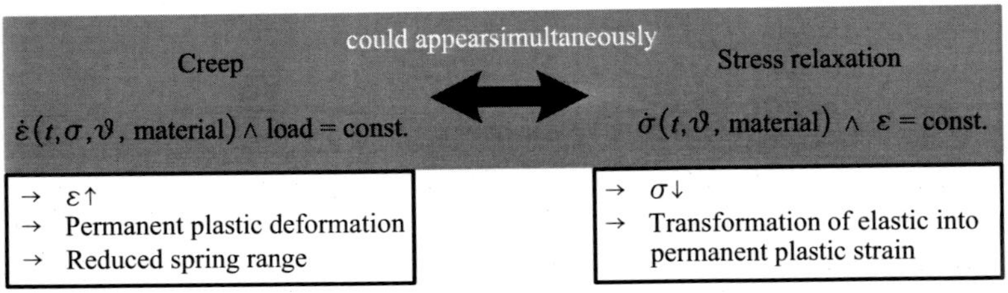

Fig. 8: Schematic relation between creep and stress relaxation

2.3 Aging of Electrical Connections and Contacts

During their lifetime, electrical connections or contacts are exposed to several kinds of stress and degradation mechanisms. These mechanisms affect the current-carrying contact area A_w and therefore change the resistance. Even small and steady reductions of the contact area can finally lead to inadmissible high resistance values. During the entire life time of electrical connections and contacts aging is present and is caused by environmental conditions and physical phenomena combined with mechanical and electrical load. All aging effects depend on time, temperature and other specific factors. For electrical connections and contacts applied in electric power systems five different degradation mechanisms are known (see also Fig. 1):

- Chemical Reactions
- Force Reduction
- Interdiffusion
- Electromigration
- Sliding Wear and Fretting (Rubbing Wear)

(Chemical Reactions, Force Reduction, Interdiffusion, Electromigration: Stationary Contacts; all five: Moving Contacts)

Chemical Reactions

Noble and non-noble metals tends more or less to develop films on the surface because of chemical reactions as oxidation or corrosion. The degradation of the contacts by chemical reactions is due the development of films. The presence of reactive elements or / and high temperatures could accelerate chemical reactions. Locally developing oxide films could grow upon the base material whereas corrosion reactions are associated with a local loss of material and the local development of corrosion products. Partially some oxidation processes are self-limiting whereas some corrosion processes could significantly reduce the material volume. All materials used in stationary electrical connections and contacts applied in electric power systems form films who have weak electrical conducting, semi-conducting or insulating electrical properties. Commonly all of them increase the contact resistance. The electrical, thermal and mechanical properties of the films all influence the degradation process of the electrical connections and contacts.

Force Reduction

A steady contact resistance requires a constant contact force. By reducing the contact force, the contact members could drift away from each other. This would result in a decrease of the number of *a*-spots and the spot radii and therefore in an increased contact resistance. The figure displayed below is valid for pure metals. For alloys further effects could occur because of the precipitation of alloying elements. Force reduction is determined through various processes and can be schematically described for pure metals as illustrated in Fig. 9.

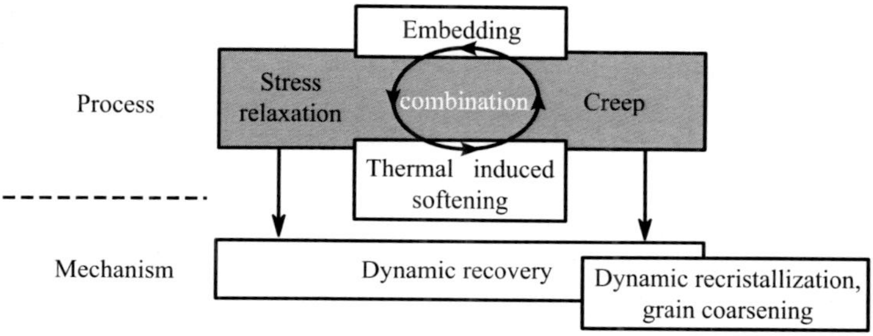

Fig. 9: Physical processes and mechanisms according to the degradation mechanism force reduction for pure metals according to [12]

The embedding process (in German: *Setzen*) takes place when the surface asperities at the contact interface will deform dependent on the mechanical load and time at the initial contact formation process (formation phase). Stress relaxation is the degradation of the tension introduced by the mechanical load and constant elongation of the entire system. This process is superimposed by the material's property to yield under mechanical load, also termed as creep (see also Fig. 8). Both processes depend on material, temperature and time. Creep is further dependent on the mechanical stress. The underlying mechanisms for both processes are dynamic recovery, dynamic recrystallization and grain coarsening [12], [13].

Interdiffusion

Because of the presence of dissimilar materials at metallic connections, intermetallic compounds could locally form at the contact interface. These layers grow between the a-spots and form an additional film resistance which could influence the constriction resistance. From its mechanical and electrical properties, the intermetallic compounds are worse than the initial contact material. The diffusion process depends on the temperature and specific material combinations [3].

Electromigration

The aging caused by electromigration depends on the presence of an electric field and its corresponding current flow which produces a resulting force and thus enforces the electrons to migrate. The larger the current density, the more effective the electromigration. The electric field creates a small electrostatic force. Usually this can be neglected when compared with the high force because of the energetic momentum transfer between negative electrons and positive metal ions. The resulting force is in the direction of the current and causes the material transport.

Although this effect can be neglected for homogenous crystal structures, it is present at the grain boundaries. Smaller grains allow for a more effective material transport then coarse grains. The diffusion along the boundaries can be divided in the following sub-diffusion mechanisms: a grain boundary diffusion, bulk diffusion and surface diffusion. Grain boundary diffusion is dominant for aluminum and surface diffusion for copper. The result of this diffusion processes is a local shift of the electron concentration and vacancies in crystal lattice. Eventually the local resistivity rises and the contact resistance increases. If the migration process persists for a sufficient time, the vacancies will exceed a critical level and voids or hillocks could occur [14], [15].

Naybour [16] discussed this effect within the field of electrical contacts as potential failure mechanism instead of common degradation due oxidation. Electromigration [17] is also presumed as a potential failure cause. Runde conducted a series of experiments on aluminum contacts as well as on the aging mechanism electromigration [18]–[22]. In combination with other degradation mechanisms, electromigration could affect the aging process and the long-term behavior of electrical contacts. In terms of contact degradation, electromigration is supposed to be much more critical, if direct current is applied. The degradation of aluminum contacts was observed for direct current (DC) and alternating current (AC) with current density $J \leq 10^6\ \mathrm{A/cm^2}$ in [23] and referred to electromigration. Hence the degradation mechanism electromigration produced by DC and AC is possibly of importance for connections and contacts in power engineering.

Sliding Wear and Fretting (Rubbing Wear)

Small relative movements of the contact members could change the contact area and therefore also the contact resistance. For example these small movements can occur through thermally dependent material expansions or mechanical vibrations induced by alternating current. Sliding wear and fretting can be distinguished by frequency and amplitude. For the most of the stationary electrical connections and contacts applied in electric power systems, sliding wear and fretting is only of minor importance and because it is more applicable to moving contacts. High power plug-in connections, typical used in gas insulated switchgear where the thermal expansion of various materials could cause sliding wear and fretting (rubbing wear).

3 Film Growth by Chemical Reactions

Adsorbed monolayers and thin films created through physical and chemical adsorption appear in the initial phase of film formation on the surface of metallic conductors. Later on, chemical reactions such as corrosion and oxidation will contribute to further film development. It can be assumed that initial films as well as later developed films influence the connections or contact resistance.

3.1 Corrosion

The term corrosion describes a local reaction between materials and their surrounding reactants. Both are defined as corrosion system. Corrosion takes place along phase boundaries and concentration differences.

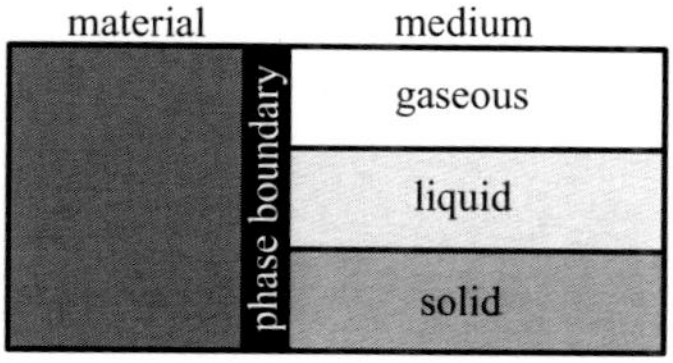

Fig. 10: Areas vulnerable to corrosion along phase boundaries according to [24]

Variables which influence the corrosion are temperature, pressure, electrochemical potential and time as well as material related properties such as alloys, grain boundaries or material composition. According to Tostmann [24] the corrosion proceeds as inner and outer corrosion. At outer corrosion, material erosion takes place at the surface. The progress of the outer corrosion is determined by the diffusion of the film's components and properties. Additionally inner corrosion can occur. This process is driven by the diffusion of elements of the corrosive environment in the material and reactions with elements in the material. Basically, three types of corrosion can occur at metallic materials: Electrochemical corrosion, chemical corrosion and metal-physic corrosion [24]. Electrochemical corrosion mostly occurs at metallic materials. Based on this several subtypes exist. They can result in various corrosion effects and corrosion damages according to DIN EN ISO 8044: 1999. More corrosion types exist which can be derived from the main types because of dominating influences, such as mechanical load or geometric properties [25].

In comparison between classical oxidation and corrosion with reference to electrical connections and contacts, many corrosion reactions are often characterized by a loss of material. This can result in significant changes in the material's structure and stability. As result force reduction could occur. This would reduce the contact area while simultaneously increasing the contact resistance. The corrosion products could further add an additional film resistance and thereby increase the contact resistance. Hence it can be stated that the result of most corrosion reactions is a combination of two following processes, film development because of chemical reactions and contact area decrease because of a loss of material and contact force. Although corrosion reactions, such as electrochemical reactions, have similar sub-processes like oxidation. Eventually they are not comparable at all because of their different impact on electrical contacts.

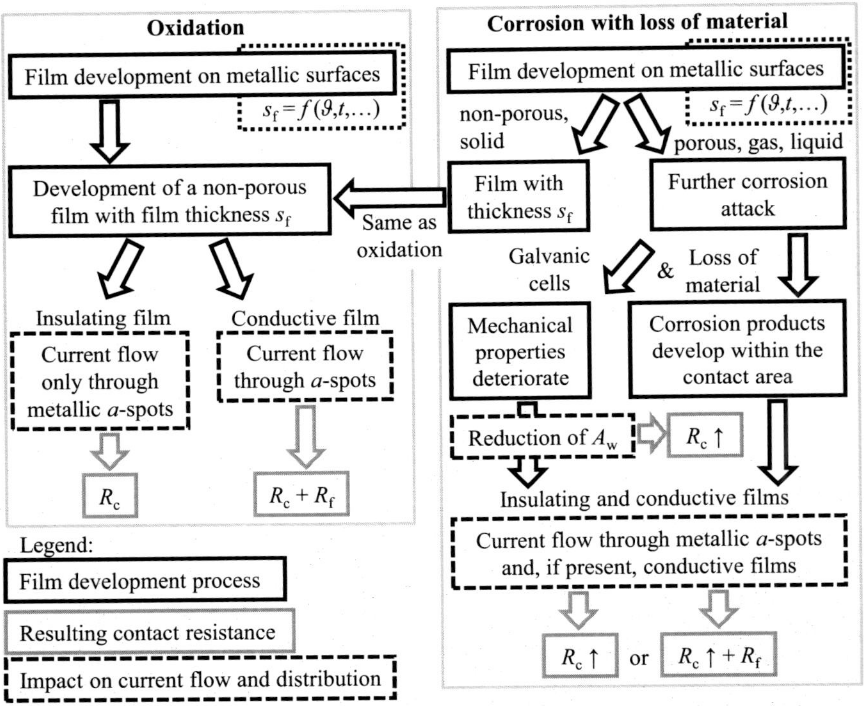

Constriction resistance R_c, film resistance R_f, true contact area A_w, film thickness s_f, time t, temperature ϑ

Fig. 11: Comparison of corrosion and oxidation impact on stationary electrical contacts with reference to film development on the surface of the contact members

3.2 Oxidation

Oxidation is commonly defined as chemical reaction which occurs together with the reduction as redox reaction. Because of a more favorable energy level, reductants will lose some electrons whereas oxidants will gain some electrons. The existence of oxygen is not a necessary condition. Basically oxidation is also a corrosion reaction. A useful technical definition established by Bürgel [11] defines oxidation as a reaction of oxygen at elevated temperatures which produces an oxide film. Initial films created by physisorption or chemisorption are assumed as already present. The growth rate of the oxide film can be described with logarithmic, parabolic, linear or breakaway oxidation relationships. Depending on the growth characteristic the film is either passivating or not. Linear Oxidation has no self-limiting growth rate and therefore no passivating film. On the other hand, the parabolic growth rate converges against a value. The self-limiting properties of an oxide can be assessed by the Pilling-Bedworth-Ratio (PBR, Appendix A)

A non-destructive ion and electron transport mechanism through the film must occur for further film growth of passivating films. Ions and electrons can pass the film through diffusion. For parabolic film growth three mechanisms can be distinguished:

a) Oxygen ions pass the film faster than metal ions. Thus, at the metal-oxide interface the film grows inwards.
b) Metal ions pass the film faster than oxygen ions. The films grows predominantly outwards at the oxide-oxygen interface. The oxide can growth unimpaired as result and no significant compressive stress occurs.
c) The mobility of the electrons is lower than that one of the ions. Depending on the diffusion of the faster ion specimen, films can both grow inward and outward.

The diffusion rate depends on the temperature, the thickness of the film and punctual lattice defects. The thermal dependency can be expressed by the Arrhenius equation. Punctual lattice defects contain of metal ion vacancies, oxygen ion vacancies and metal or oxygen interstitials. The predominant lattice defect determines the diffusion rate and therefore the film growth. All defects influence the quantities of the reactants in the possible reaction, the stoichiometry. E.g. for a given oxide with an impaired stoichiometry ratio and oxygen shortage the film growth will be determined by oxygen diffusion. Oxide films with balanced stoichiometry rations have low defects and therefore enable only low diffusion velocities [11]. Stable oxides do have a high homolog temperature. The several processes are illustrated in the figure below.

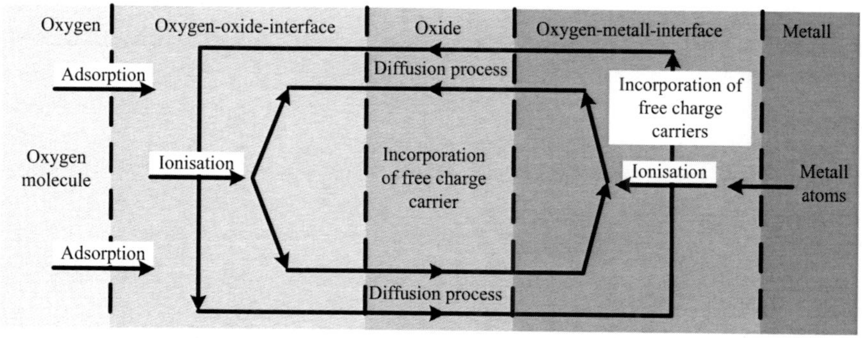

Fig. 12: Schematic process of oxidation with movement, exchange and incorporation of charge carriers according to [4]

Oxidation of Alloys

In addition to the explanations above further effects have to be considered for alloys. The various elements differ in oxygen affinity, material concentration, diffusion rate and in the composition of the developed surface film. Based on the elements properties described above, two oxidation mechanisms can be distinguished: selective oxidation and inner oxidation.

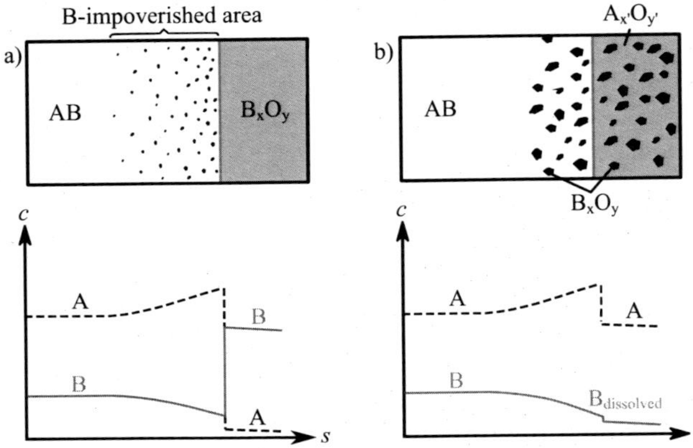

Fig. 13: Schematic Oxidation within an alloy with main element A (noble) and alloy element B and concentration *c* across thickness *s*; in a) B develops a close film B_xO_y, the area below the film impoverish of B, only a few elements a are dissolved in the oxide film; in b) the concentration of B is too low to develop a close film, B oxide will be incorporated in the $A_{x'}O_{y'}$ film and a few B elements are dissolved. Figure according to [11].

Non-porous film develops because of sufficient high concentration (molar fraction ≥ 10 % according to [11]) and better thermodynamic properties of element B in Fig. 13. Thus alloy depletion results. If the element concentration reaches below a critical value in order to create a non-porous film, inner oxidation could occur. In this case Element A dominates the oxide growth and develop a film. Depending on the alloying element's properties and partial pressure of oxygen, B elements oxidize and will be integrated to the A-oxide. The oxidation within the material is defined as inner oxidation. Both mechanisms can interfere. Three different states of alloy oxides can be specified:

- No Solubility of the corresponding oxides
- Mixed crystals
- Different oxide types such as twin-oxides

3.3 Films on the Surface of Stationary Electrical Connections and Contacts in Electric Power Systems

In power engineering mostly pure copper and pure aluminum as well as selected aluminum alloys serve as base materials whereas silver, tin and nickel being coatings dedicated to specific applications.

3.3.1 Copper

Electrolytic-Tough-Pitch copper (Cu-ETP) or oxygen-free copper (OFC) is the mostly used copper in electric power systems and it is characterized by a high purity. On the surface of copper and in atmospheric air copper(I) oxide Cu_2O develops at temperatures up to 260 °C. At higher temperatures copper(II) oxide CuO develops and covers the Cu_2O film [3], [25], [26]. In marine atmosphere the corrosive attack is stronger due to chlorine and ammonia.

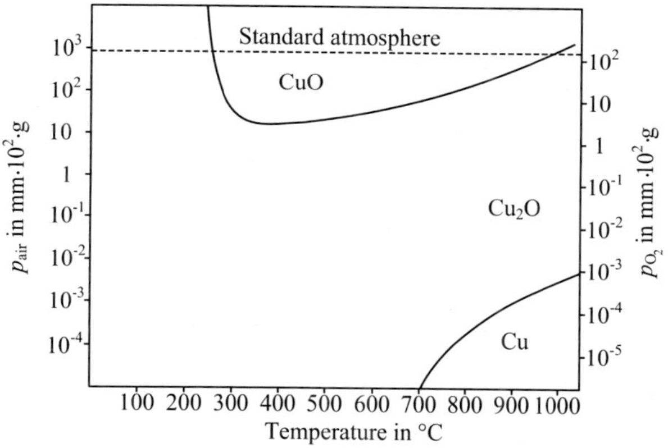

Fig. 14: Development of Cu_20 and CuO dependent on air pressure and temperature according to [26]

Copper(I) Oxide Cu_2O - Physical Properties

Cu_2O has a monoclinic crystal system. Similar to copper itself the copper atoms in Cu_2O are cubic disposed. The oxide has therefore good adhesive properties on copper [27], [28]. The length of one unit cell of Cu_2O is 0.42696 nm and is therefore ≈ 18 % larger than copper itself. The volume of a Cu_2O unit cell is around 65 % larger than that one of pure copper. During the oxidation no restructuring of the lattice takes place but rather a lattice widening [29]. For Cu_2O a Vickers hardness of 109 to 189 HV was determined [27], [28], [30]. For Cu-ETP a Vickers hardness of 95 HV was determined [31]. Copper(I) oxide is harder than copper itself. The color of pure Cu_2O depends on the temperature, time and thickness. Depending on the grain size, pure Cu_2O appears as yellow, orange or red. Thin films appear to be more transparent; thick films appear as brown. More thick films developed at higher temperatures initially appears as red, then bronze, then brown and finally merging into black [27], [28], [32].

Copper(I) oxide is a p-conductor by vacancies as electron holes. The stoichiometric Cu_2O lattice consists of monovalent positive copper ions and bivalent negative oxygen ions. Point defects, such as vacancies are in the same order for positive and negative charge carriers [33]. New oxides develop at the interface air-oxygen and positive copper ions migrate from the oxide lattice (Fig. 12). In order to reach a balanced charge level bivalent positive copper ions develop in the oxide lattice. The oxide's conductivity depends on the concentration of vacancies and on specific development conditions, such as oxygen partial pressure [25], [28]. The oxide's properties are sensitive to minor changes of the conditions in the oxide development process. Already slightly variations can result in several magnitudes difference for the resistivity. For power engineering naturally developed oxide is of most interest. Slade expresses a temperature-dependent equation with the Boltzmann constant k_B and an 0.3 eV activation energy [3]:

$$\rho_{Cu_2O} = 0.38e^{\frac{0.3eV}{k_B T}} \tag{25}$$

Tab. 3: Specific electric resistance of Cu_2O at room temperature [27], [28], [31]

	ρ_{20} in Ωm
Cu ETP	$1.8 \cdot 10^{-8}$
Synthetic created Cu_2O	$5 \cdot 10^5 - 3.7 \cdot 10^7$
Synthetic created Cu_2O of highest conductivity	$4.5 \cdot 10^{-1} - 1 \cdot 10^1$

Copper(II) Oxide CuO - Physical Properties

Copper(II) oxide CuO has a monoclinic lattice. The lattice of CuO is base-centered. The copper ions of CuO are not cubic disposed so that lattice restructuring of Cu_2O to CuO occurs during the oxide development [27]–[30]. CuO has less adherent properties on copper due to the different lattice structure [27], [28]. The volume of a CuO unit cell is about 73 % larger compared to copper. Copper(II) oxide is harder than pure copper and optically appears as iron-black, steel-gray or bluish. According to Fig. 14 CuO only develop at temperatures above 260 °C. CuO is a semiconductor with electrons and holes serving as charge carriers. The conductivity of CuO is slightly better than that for Cu_2O (Appendix A, Fig. 74).

Film Growth

A bare copper surface would oxides quickly at room temperature due to the oxygen adsorption [1]. Molecular oxygen will be dissociated and bonded to free copper atoms at the metallic surface. A monolayer of oxide appears after a few seconds [1]. A closed film appears after a few minutes and the growth rate will be reduced [26], [34], [35]. Growth ratios for time spans of days or weeks can be found in the literature which are partially not consistent. However, all research confirms decreasing growth ratios at a constant temperature. The further film development is particularly determined by diffusion of copper and oxygen ions. It depends on the temperature [25], [36]. The structure of the crystalline lattice of the copper has also an influence on the oxide film growth and can be described by the Miller indices. Lower Miller indices have higher growth ratios for copper [26]. The concentration difference at the oxide - metal interface with more copper ions on the metal side and more oxygen ions on the oxide side results in a electrochemical potential across the oxide layer. As a result, the free energy within the crystals is increased and initiates an ion transport to rebalance the concentration difference within the mixed crystal (Fig. 12) [33], [35]. In crystalline solids the diffusion takes place across vacancies, interstitials and grain boundaries. It is dominated by the crystal system. In Cu_2O diffusion across vacancies is presumed [4], [33], [36]. The diffusion process can be described mathematically by Ficks law of diffusion. The solution for diffusion across vacancies in crystalline solids can be stated:

$$D = D_0 \cdot e^{-\frac{Q_a}{R' \cdot T}} \tag{26}$$

At normal pressure and normal temperatures up to 200 °C, the frequency factor D_0 for copper is $2 \cdot 10^{-5}$ m² · s⁻¹ and for oxygen $6.5 \cdot 10^{-7}$ m²· s⁻¹ [33], [36], [37]. According to [36], [38] the activation energy Q_a for copper and oxygen is similar and is in the range 140 – 160 kJ · mol⁻¹. R' is the gas constant. When comparing the diffusion coefficient of copper and oxygen it is notable that the mass transport of copper ions is more than thirty times larger. Other research observes up to fiftyfold [36]. Hence for common oxidation processes, the oxygen ion diffusion can be neglected whereas the copper ion transport determines the rate for oxide growth [25], [26], [33]. Holm [1] states an equation for the film thickness s in nm, the time t in h and the Boltzmann constant k_B. This is also more known as parabolic law:

$$\frac{s}{\mathrm{nm}} = \sqrt{4 + 7.94 \cdot 10^{12} \cdot e^{-\frac{1eV}{k_B \cdot T}} \cdot \left(\frac{t}{h}\right)} \tag{27}$$

Vernon confirms the parabolic behavior, which only differs in a higher measured film thickness [39], [40]. A high relative humidity slightly decreases the growth rate [39]–[41]. If hydrogen sulfide H_2S is present, a dark film will develop rapidly. The film mainly consists of Cu_2O but is not protective [27], [28].

3.3.2 Aluminum

Besides copper aluminum is the most important conductor material in electric power systems. In order to increase the mechanical properties of aluminum it is partially alloyed with magnesium and silicon. Under normal conditions an amorphous Al_2O_3 oxide film develops rapidly on pure aluminum surfaces. If humidity is present, a further more porous cover layer develops on the Al_2O_3 film (Fig. 15) [42]. The water containing cover layer will be penetrated either because of the high force of electrical contacts in electric power systems or decomposes thermally because of the elevated temperature during operation. Impurities on the material surface will be incorporated in the film that appear as mixed oxide. Alternatively intermetallic compounds with the aluminum will develop [43].

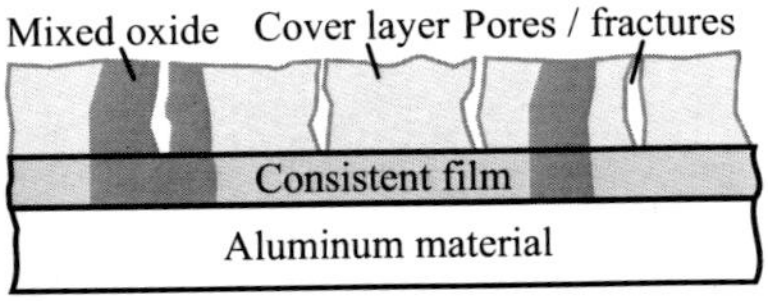

Fig. 15: Composition of aluminum oxide films according to [42]

Aluminum Oxide Al_2O_3 - Physical Properties

Al_2O_3 is a polymorph material with a stable α-modification. α-Al_2O_3 is known as corundum with sapphire or ruby. Corundum has an trigonal rhombohedral lattice with ten atoms per unit cell [44]. Aluminum oxide is far harder than aluminum, partially by a factor of up to 100-fold [30], [44]. Up to temperatures of 400 °C the oxide is amorphous [45], [46]. It is questionable whether the properties of crystalline oxide can be transferred to the amorphous oxide. Several sources describe the oxide as very hard, well adherent and brittle [1], [3], [7], [44]. If the oxide is fractured the cracks influence the size and shape of the a-spots. The thermal expansion coefficient $\alpha_{th,Al2O3} = 6.2 \cdot 10^{-6} \cdot K^{-1}$ is essential smaller than that of aluminum $\alpha_{th,Al} = 24.5 \cdot 10^{-6} \cdot K^{-1}$. High mechanical stress could develop for strong temperature changes. Optical aluminum oxide films appear as transparent and can therefore not be detected manually.

Within aluminum oxide the conduction occurs across electrons. The resistivity of Al_2O_3 is above $10^{12}\ \Omega m$ at common operating conditions of 90 °C in power engineering and significantly larger than pure aluminum (Appendix A, Fig. 74).

Film Growth

The initial oxidation of a bare aluminum surface occurs instantly after cleaning. After a few seconds (≈ 20 s) seven layers of molecular oxygen are rapidly adsorbed [1]. Within a few minutes a close film is developed. The film growth slows significantly down before it finally stops. Thickness and composition depend on the ambient temperature and the material composition. The final thickness of Al_2O_3 and cover layers depends on both the temperature and the relative humidity. The cause for the limited film growth is presumed to be the lack of crystal defects in Al_2O_3 film. As a result no diffusion across the oxide film occurs. In order to transport reactants across the film an electric field is necessary. This field develops due to the high concentration surplus of negative oxygen ions at the oxide-air interface and positive metal ions on the metal-oxide interface. The aluminum ions are half the size compared with the oxygen ions. Thus the dominating diffusion process is the outward transport of the aluminum ions [47]. The electric field effect is not strong enough to maintain the film growth above a specific film thickness [48]. Aluminum surfaces need a thorough cleaning to ensure stable electrical contacts.

Influence of Alloying Elements

Alloying elements influence the composition and the growth of the oxide. Magnesium is the most important element due to its high oxygen affinity. According to [49], [50] MgO (Tab. 4)

Tab. 4: Oxide reaction products of aluminum alloys depending on the Mg mass concentration

Mg content in m%	> 1	$10^{-4} - 1$	$< 10^{-4}$
Stable reaction product	MgO	$MgAl_2O_4$	γ-Al_2O_3

develops at room temperature and will be incorporated within the oxide. The diffusion affinity of Mg from the metallic material into the oxide towards the oxygen-air interface increases with rising temperature. Pure Mg is enriched at the metal-oxide interface [51]. In [50] an overview of literature for research on aluminum alloys with Mg contents $\leq 1 \wedge > 1$ m% is presented. No experimental research investigating the typical temperature of electrical contacts and connection in electric power systems was found. However, it is assumed MgO could appear in the oxide film of aluminum alloys at higher temperatures. The additional alloy silicon only as has a minor influence compared to magnesium. Its presence in the film is also enhanced at elevated temperatures [51].

3.3.3 Silver

Silver does not form an oxygen based oxide at common conditions. When ozone is present on the silver surface, Ag_2O can develop. Ag_2O is a very soft oxide with only weak adhesive properties. Due to that it can be easily removed by mechanical treatment or high forces. Ag_2O began to decompose at 200 °C. For electrical connections and contacts Ag_2O is usually not important [1], [7]. Silver sulfide Ag_2S is developed by a chemical reaction of the silver surface with hydrogen sulfide. Sulfur hereby oxides with the silver and silver sulfide develops. Ag_2S is considered as a polymorph material at normal pressure. It can be classified as ascendant according to the temperature areas of α, β and γ modifications. The change between the modifications takes place at 179 °C and 586 °C. Thus the α and β modifications are of interest for stationary electrical connections and contacts in electric power systems.

Silver Sulfide Ag_2S - Physical Properties

The α-modification of silver sulfide is monoclinic whereas sulfide ions are body-centered cubic. The β-modification has cubic lattice, as well as the sulfide ions The silver ions are distributed statistical [52], [53]. Naturally α-Ag_2S appears as acanthite and is softer than pure silver itself. The hardness of Ag_2S is one third of the silver hardness [54]. Hence a possible silver sulfide layer would be pushed away before the silver begins to creep. Ag_2S appears as a lead-gray, gay or iron-black color [55].

The conductivity of the α-modifications enlarges with an increasing temperature until the change to the β-modification takes place. The electron conduction in the α-modifications is driven by positive silver ions and electrons, whereas the conductivity of the β-modifications predominantly relies on electrons and is therefore higher (Appendix A, Fig. 74). With temperature increase, the resistivity of Ag_2S continually decreases. For temperatures higher than the transition point, the electrical behavior of Ag_2S significantly improves from

α-modification to β-modifications at $\vartheta \approx 179$ °C. Nevertheless the conductivity of silver is roughly more than three magnitudes above that of Ag_2S. The growth of silver sulfide is determined by the presence of hydrogen sulfide H_2S [53]. The atmosphere contains only a few parts per million of H_2S in common environments. The growth process is slow, compared with aluminum. At the surface Ag_2S local volumes with radii of (2.5…6.25) nm develops as non-closed film. The growth rate will get constant after approximately ten days and can then be expressed by a linear growth rate. As the Ag_2S film growth is not passivating but rather continues, this process is therefore not dominated by diffusion. Silver ions are also characterized by a high mobility [53]. The growth rate is approximately determined by the slow adsorption of new sulfur elements which in turn is influenced by the presence of H_2S. The gas circulation within the contact area is still lower and thus indicates a lowered growth rate [56].

3.3.4 Tin

The metal tin is often used as coating. Tin has a lower melting point and limited thermal operating range. Compared to other metals, tin can be easily finished mechanically. At temperatures till the melting point of tin the metastable Tin(II) oxide SnO [57]–[60] can occur. In a temperature area of 175 °C - 300 °C, a disproportionation takes place and Tin(II) oxide will be reduced to tin and tin(IV) oxide [57], [61], [62]: $2\ SnO \rightarrow SnO_2 + Sn$. As a result, tin(II) oxide converts intermittently into Sn_2O_3 and Sn_3O_4 and finally in tin dioxide SnO_2 and metallic tin.

Tin(II) Oxide SnO and Tin(IV) Oxide SnO_2 - Physical Properties

Although much harder than tin itself, tin(II) oxide is a soft material It has a tetragonal crystal system. Naturally SnO appears as mineral romarchite [63]. The color of SnO ranges from blue to black, thin layers could also appear in other colors. After being exposed to an indoor atmosphere for 150 days, the oxide appears as slightly yellow-gray. Through an enlarged viewing angle the oxide appears more bluish-gray [64]. SnO_2 has also a tetragonal crystal system and occurs naturally as cassiterite. It is harder than SnO and pure tin [57], [65]. SnO_2 is mostly transparent and appears as yellow or cinnamon-brownish [62].

At room temperature tin(II) oxide is a p-type semiconductor with an additional electron hole. The resistivity of tin(II) oxide strongly depends on the development process and its boundary conditions [49], [62]. The mode of conduction changes to electron conduction at 200 °C. A relation between disproportionation and the beginning transformation of SnO in SnO_2 might be possible [62]. With increasing temperature the resistivity of SnO decreases (Appendix A, Fig. 74). SnO_2 is also a semiconductor with a low resistivity that is less dependent on temperature (Appendix A, Fig. 74).

Film Growth

The initial slow oxidation phase of tin at different oxygen pressures was observed in various investigations [58]–[60]. No fundamental dependency between oxygen pressure and growth rate and process was observed. Both processes are assumed to be identical at the same standard conditions. Initially developing oxide particles form the center of circular growing oxide films at the surface. These centers have a temperature independent density of $\approx 2 \cdot 10^6$ cm^{-1}. It is assumed that the starting points of these centers are lattice defects. From the centers the films then grow in lateral direction until the whole surface is covered. Afterwards the film begins to grow in vertical direction perpendicular to the surface. Within a third of the film between tin and tin(II) oxide small voids develop (Fig. 16).

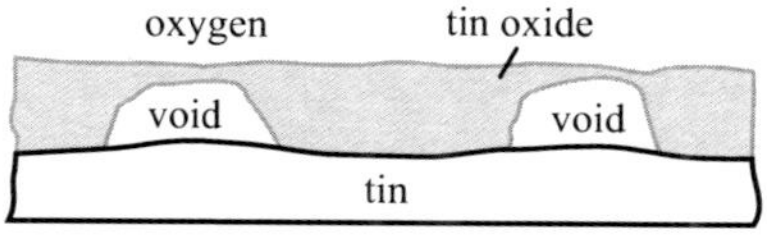

Fig. 16: Schematic structure of tin(II) oxide film on tin surface and oxygen atmosphere according to [12]

These voids are only covered with a thin layer of tin(II) oxide. Due to the ongoing film growth the mechanical stress in the thin cover layer above the voids increases until a fracture occurs. Oxygen reaches the tin through the now partially open voids and the oxidation process restarts until the void is filled up with oxide. The film becomes fragmented over time. When the surface is not fully covered during the initial phase of growth, the growth depends on the temperature. If the surface is fully covered with oxide the growth rate is logarithmically and is decreasing until the film becomes fragmented because of the voids. Then the growth rate increases again. According to long-term investigations a nearly linear growth ratio was determined [64]. Therefore, the film is assumed not to be passivating. In humid and circulating air the oxidation rate is assumed to be higher.

3.3.5 Nickel

Together with oxygen, nickel forms nickel(II) oxide NiO. NiO naturally appears as a rare mineral bunsenite. Nickel(II) oxide is a very hard, toxic and non-flammable material. Due to the high oxide hardness nickel is less useful for switching contacts. Electrical NiO is highly resistive and insulating. In harsh environments nickel combined with NiO films better withstand corrosion processes than other metallic materials in power engineering. In comparison to other films the growth rate is much lower [1], [33], [66]. Holm states the following formula for a film growth with the thickness s in nm being dependent on the temperature ϑ and the time t at standard conditions under common laboratory atmosphere [66].

$$\frac{s}{\text{nm}} = 0.0484 \cdot e^{0.0137\left(\frac{\vartheta}{°C}\right)} \cdot \ln\left[25\left(\frac{t}{h}\right)\right] + 2 \qquad (28)$$

It is assumed that hereby a (2…3) nm thick film already exists due to initial physical and chemical adsorption processes. The formula was confirmed under experimental conditions. At varying environmental conditions the growth could change. For higher levels of relative humidity the growth rate would increase [1], [67].

4 Prior Works on the Long-Term Behavior of Stationary Electrical Connections and its Degradation Induced by Film Development due to Chemical Reactions

Stationary electrical connections and contacts applied in electric power systems have to fulfill high reliability requirements in order to ensure a continuous and stable operation. Generally speaking, technical devices used for the generation, transmission and distribution of electrical energy are permanently in operation for long periods of several thousand hours. The long-term behavior is therefore crucial for the performance of such connections and contacts. In total the operating time of stationary electrical connections and contacts in electric power systems can be several tens of thousands hours. In contrast to electrical connections and contacts, e.g. automotive applications, stationary electrical connections and contacts in electric power systems obtain a continuous energy input during their lifetime whiteout long operational interruptions. All degradation mechanisms somehow depend on temperature and time. Significant changes of these parameters can ultimately alter the underlying physical processes. For experimental investigations on electrical connections and contacts applied in electric power system, these operational parameters have to be considered carefully in order to obtain reliable information about the long-term behavior.

The degradation of electrical connections and contacts based on chemical reactions is supposed to be more aggressive if the surrounding environmental conditions promote chemical reactions. This would indicate that the harsher the environments, the worse the connection or contact performance. Some standards address the relationship between the environment and electrical devices. The national edition of the International Electrotechnical Commission (IEC) standard IEC 60721-1 is the DIN EN 60721-1: 1990 + A1: 1992 + A2: 1995 "Classification of environmental conditions". The standard defines certain classes for different environmental conditions and also covers various existing climates worldwide. The IEC 60068 or DIN EN 60068-1: 1994 + Corrigendum 1988 + A1: 1992 "Environmental Testing" on the other hand, determines environmental testing with reference to IEC 60721. According to IEC 60721-1: 1990 + A1: 1992 + A2: 1995 environmental conditions are characterized by one or more environmental factors which in turn are characterized by environmental parameters (Fig. 17). For example the environmental condition "desert climate" is characterized by heat which can be influenced by further environmental parameters such as radiation, ambient temperature and other.

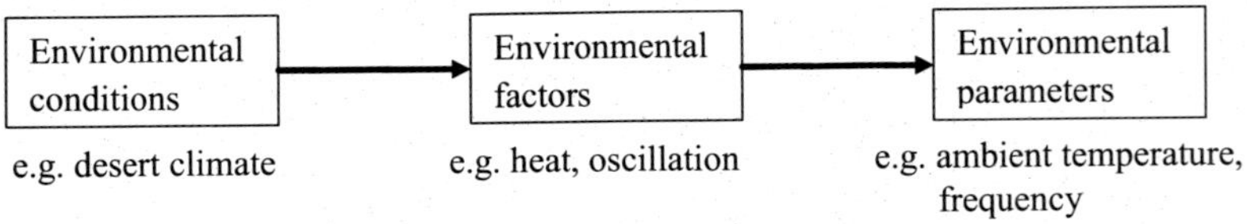

Fig. 17: Structure of environmental definitions according to IEC 60721-1

A few subchapters in IEC 60068 are designated to electrical contacts. Although they outline a selection of specific environmental conditions which could occur in operational services, not all possible operating conditions are addressed. Commonly the test procedures of IEC 60068 are designated for testing of specific electrical devices and not for long-term test of electrical connections and contacts. In field electrical connections and contacts of different sizes and applications could be stressed by the same environmental parameter in different ranges. Therefore, a solely limitation to an environmental parameters range would only match specific but not overall operating conditions.

In field, and with respect to the aging by chemical reactions, a few environmental parameters and factors can be identified which affect the long-term behavior of electrical connections and contacts. These parameters include: ambient temperature ϑ_{amb}, ambient temperature spread $\Delta\vartheta_{amb}$, relative humidity RH and reaction accelerators (gas components). The ambient temperature ϑ_{amb} limits the current load while high temperatures accelerate aging processes. Lower temperatures on the other hand allows for higher loads and thus current levels. The ambient temperature spread ΔT_{amb} between the highest and the lowest temperature value can be important if its range leads to a significant material expansion. High rates of relative humidity RH accelerate the aging process due to a higher corrosion rate [68]. The reaction process always depends on the participating reactants. Reaction accelerators in the air include sulfides, sulfates, chlorides, nitrates, hydroxides, or ammonium salts [68]. If they occur individually, their possible influence would be much less critical than in combination [69]. The corrosion rate is the highest in combination with increased relative humidity.

Prior Works on the Long-Term Behavior of Stationary Power Connections Applied in Electric Power Systems with Reference to the Degradation Mechanism Chemical Reactions

In the mid 20th century the increased use of aluminum connectors in electric energy infrastructures in harsh environmental conditions has led to more failures and therefore to initial research towards the long-term behavior with respect to the environmental conditions. Sander's initial research purpose [70] was to evaluate the performance of the given aluminum connectors as well as to offer information for the further development of future fittings. Early findings were conducted by field and laboratory research. These studies investigated the performance of electrical connections in a marine environment with load cycling and different types of conductor's. Unfortunately they lack an evaluation of the contact resistance with reference to the degradation mechanisms. First work were done by Bonewitt and Sander. In laboratory and field tests, they studied the performance of typical power connections exposed to a marine environment [70], [71]. The already noted the difficulties when connecting aluminum and copper and have focused on corrosion degradation. Basic research work on the principle of chemical reactions, especially corrosion, was done by Antler and Abbott. They defined standard test environments dependent on different gas components as well as made some basic conclusions about the assumed interaction between film and electrical contact [68], [72]–[74].

However, both did not focus on the long-term behavior of stationary electrical connections applied in electric power systems. Less known internationally are the investigations of Grossmann [75] who did research on high power plug-in connections applied in electric power systems. He studied the long-term behavior of high power plug-in connections dependent on thermal, mechanical and environmental parameters as well as design and material. Contacts with a higher initial resistance undergo a faster aging process. The aging process occurs faster for opening contacts and contacts with higher temperatures. It is stated that the resistance $R(t)$ depends on the initial resistance $R(t = 0)$ and that the aging rate $\mathrm{d}R / \mathrm{d}t$ depends on the given resistance $R(t)$. The rise of $R(t)$ was reasoned by film ingrowth dependent on environmental conditions and the thermal load at the contact area [75]. According to this, the long-term behavior of electrical connections and contacts is determined by initial and operational conditions. With $\mathrm{d}R / \mathrm{d}t \sim R(t)$ and a constant temperature, the aging of electrical connections and contacts can be described by an aging model. This model was established by Böhme / Löbl [76] and later extended by Grossmann [75], [77].

At stationary connections and contacts the frequent measurement of the connection or contact resistance allows to evaluate the status of the connection or contact dependent on time. The connection or contact resistance directly influences the connection or contact temperature. For technical connections such as bolted busbar joints, the performance factor k_{u} can be used to evaluate the connections performance as a ratio between the resistance of the bolted busbar joint R_{J} and a busbar sequence R_{busbar} of same length [78]:

$$k_{\mathrm{u}} = \frac{R_{\mathrm{J}}}{R_{\mathrm{busbar}}} = \frac{U_{\mathrm{J}}}{U_{\mathrm{busbar}}} = \frac{P_{\mathrm{J}}}{P_{\mathrm{busbar}}} \qquad (29)$$

The lower the factor, the better the electrical and thermal performance of the connection. In a later study by Bergman [79], the model introduced by Böhme / Löbl was expanded to now include experimental results on the long-term behavior of bolted busbar joints. The experimental study [79] of aluminum, copper and aluminum-copper connections were carried out with reference to different cleaning processes, shapes, hole patterns and environmental conditions. The expanded aging model in [79] depends on the temperature, the passivating effect of films and the transition of the current field and determines the time dependent performance factor $k_{\mathrm{u}}(t)$.

The application of this model for different shapes is questionable because of the busbar design of the experimental study. Furthermore, the interactions between current field and films seem to be neglected. However, for both considered studies [75], [79] it was stated that the contact and connection degradation in the underlying experiments is dominated by films because of chemical reactions. Bergmann mentioned that during the formation phase the chemical reactions at the circumference of the a-spots and the environment continues unimpaired. The process result in decreasing a-spot size until the film reaches a self-limiting thickness which is dominated by a slow diffusion process. This hypothesis cannot be confirmed for all shapes and materials of connections and contacts in other research [3], [7]. Bergmann further states that the resistance slowly increases over time until a steep increase of the resistance appears during the phase of accelerated aging.

More recent work by Callen and Johnson et al focuses on the effects that a marine environment can have on the performance of power connectors, particularly aluminum-bolted compression and fired wedge connectors [80], [81]. Long-term studies about several ten thousands hours were conducted in a marine environment at ambient temperatures. It was found that the loss of mechanical contact load at the conductor-connector interfaces because of corrosion and corrosion-product buildup, was a major factor controlling connector life. Fairley stable performance of some connectors in this work [80], [81] mainly came from the "spring-action" of some connectors which resists the loss of mechanical contact. Further reasons were a more uniform stress distribution over the contact interfaces and the larger area of metal-to-metal contacts. It was shown that the usage of inhibitors in mitigating corrosion was rather limited in all cases [80]. Considering the result of [80], [81], it seems that the separation of the impact of force reduction and film development by corrosion on the connection resistance is difficult. More recent research on the true influence of films by chemical reactions was done by Takano [82]. He studied the degradation of copper contacts made of crossed rods at very low forces of 1 N up to 10 N at a temperature of 200 °C. He observed a significant contact resistance increase after a few 100 hours. Distinct oxide films were observed on the surface of the used rods by microscopy. Takano assigned the oxide formation as cause for the contact failure. However, Takano's results indicate no degradation for test specimen with 5 N and 10 N contact force. Additional research work on the long-term behavior of bolted busbar joints was done by Schlegel [12]. He studied the long-term behavior of different coating systems on copper with respect to the degradation mechanisms force reduction, interdiffusion and particularly chemical reactions. It was found that copper oxide films form a closed non-porous layer which would result in a very slow ingrowth of the films into the contact area [12]. In his long-term study over 20000 hours, Schneider [83] did tests on bolted busbar joints with aluminum-silver and aluminum-copper interfaces and focused on the degradation mechanisms force reduction and interdiffusion. He observed some unexpected resistance changes in long-term studies on bolted busbar joints in oven filled with gaseous nitrogen at 200 °C. He assumed that the observed degradation is not solely caused by Interdiffusion and stated that possibly oxide films, produced by chemical reactions, could be present [83].

Existing aging models often explain the performance deterioration with chemical reactions and thereby neglect other degradation mechanisms such as force reduction. Because of technical applications, different kinds of connection geometry are often used for experimental research. This partially prohibits the comparability of the results. Major differences amongst the experimental designs as described above also exists in terms of operating time. Further notable are variety in measurement devices as well as measuring techniques.

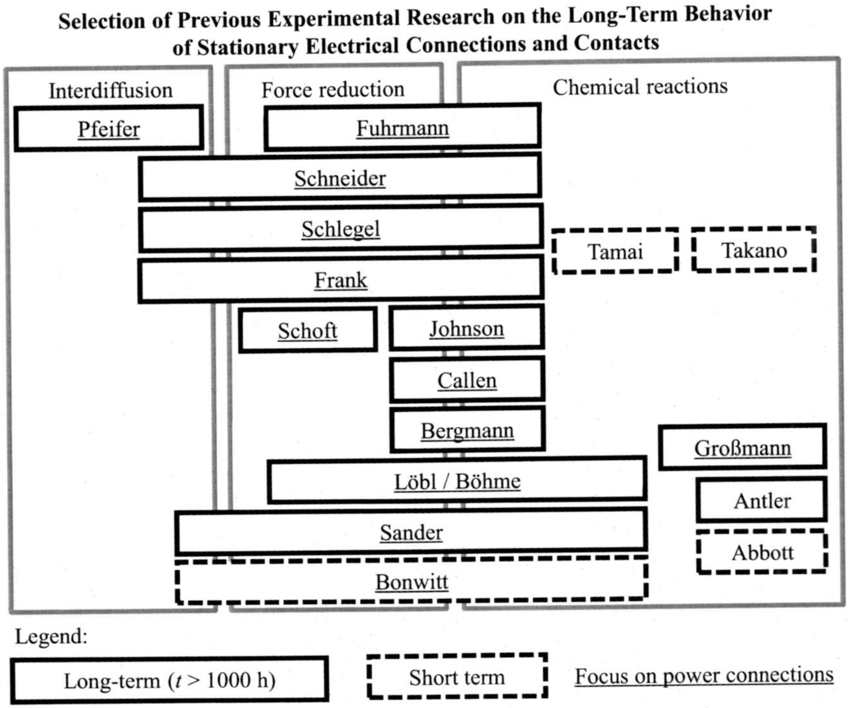

Fig. 18: Overview of previous investigations with respect to other aging mechanisms, experimental test duration and power connections [12], [68]–[71], [73]–[76], [79]–[90]

Different degradation mechanisms in the considered research (Fig. 18) are not always distinguishable from each other or cannot be separated with reference to film growth by chemical reactions because of:

- Connection and contact design
- Temperature
- Measurement technique
- Physical mechanism
- Experimental testing time

The following applies for many of the studies mentioned above and also for further studies. Different degradation mechanisms could be the cause for enhanced contact resistance but cannot be delineated from each other with respect to the influence of films because of chemical reactions. Practical investigations are often more applicable to reality and neglect to provide information on the physical mechanism. No convincing experimental data on the long-term behavior of stationary electrical connections and contacts in areas affected by natural harsh environmental exist with exception of [70], [75], [79]–[81]. Nevertheless, they somehow lack the opportunity to conclude on the true influence of the degradation mechanism chemical reactions on stationary electrical contacts.

5 Definition of Tasks

The film development by chemical reactions is often stated as cause for the degradation of stationary electrical connections and contacts. However, no distinct evidence was found in the literature. Considering the true impact on the long-term behavior, more clarification is needed. The scope of this work is to study the impact of films by the degradation mechanism chemical reactions on the long-term behavior of stationary electrical connections and contacts applied in electric power systems. Research on circular contact points will be conducted while suppressing other degradation mechanisms than film development caused by chemical reactions. This work will focus on film development without significant loss of material and contact force (oxidation).

Moreover, the conducive environmental influence on the film development is known. No data can be found in the literature considering the long-term behavior of stationary electrical connections exposed to harsh natural environments such as desert and tropical rainforest environments. The long-term behavior of bolted busbar joints will be studied in long-term tests with reference to the surrounding natural environment.

The tasks below can be deduced:

- Designing of an experimental setup to study the individual degradation mechanism film development by chemical reactions and suppression of other degradation mechanisms
- Studies on the initial state of stationary electrical contacts such as material properties, initial contact resistance and film influence
- Determination of the mechanical stress in the contact area at initial conditions
- Microscopic analysis of the contact area at initial and aged condition
- Studying of the influence of initial film development in the contact formation phase
- Determination of the physical interactions between film growth and contact resistance depending on time and temperature
- Designing and construction of experimental tests to study the long-term behavior of stationary electrical contacts applied in electric power systems with reference to the environment, the material and the temperature
- Evaluation of the film's impact on the constriction and the contact resistance
- Designing and construction of an experimental setup on the long-term behavior of power connections such as bolted busbar joints operated in harsh environments as well as long-term tests with the designed setup dependent on time, temperature, material and environment

Within this framework, the impact of film development on the long-term behavior of stationary electrical connections and contacts will be evaluated with reference to the contacts degradation.

6 Impact of Film Development on the Long-Term Behavior of Stationary Electrical Connections and Contacts

In power engineering, copper and aluminum are common materials for those connections and contacts. The studied bulk materials are electrolytic tough pitch copper Cu-ETP, pure aluminum Al and an aluminum alloy AlMgSi. Cu-ETP is a standard material found in power engineering and is regularly used in high current components. Aluminum and its alloys are commonly used materials for transmission line components and substations. The contact surface can be improved by coatings. Typical coatings in power engineering, such as silver, tin, nickel were coated on copper for this research (Tab. 5).

Tab. 5: Studied material combinations

Bulk material	Cu-ETP				Al99.5	AlMgSi0.5
Material number	CW004A R250				EN AW-1050A H111	EN AW-6060 T6
Material system		Coating				
	Bare	Sn (16 μm) Ni (1 μm)[2]	Ag (16 μm)	Ni (16 μm)	Bare	Bare
Designation	Cu	Cu-Ni-Sn[2]	Cu-Ag	Cu-Ni	Al	AlMgSi

[1] Contacts are termed by material combination and slash " / " at the interface area, e.g. pure Cu contacts → Cu / Cu; substrate and coating are associated with a hyphen " - ", e.g. Cu-Ni

[2] Nickel is plated as interlayer between Cu and Sn

6.1 Material Properties

Chemical Constitution Substrate

The chemical composition of the investigated substrates was controlled by inductively coupled plasma optical emission spectrometry (ICP-OES) and thermal extraction with a carrier gas (TECG). No deviations from the standards were detected (Appendix B, Tab. 23).

Coating System

Coatings are frequently used in power engineering to improve a contact system's thermal and electrical properties, corrosion resistance as well as other characteristics. For this study several copper rods were coated with silver, tin or nickel. The coating's thickness was specified in order to prevent corrosion through the coating layer. Thin coatings can lead to corrosion between substrate and the environment through the coating [91]. Therefore, a thickness of ≥ 16 μm for tin, silver and nickel was utilized. This process is also displayed below (Tab. 6).

Moreover, chemical bonds could develop between several firmly bonded metals. At increased temperature, the material combination of tin-copper tends to create high resistive intermetallic compounds. A solid overview of the intermetallic layers between Cu and Sn regarding stationary electrical connections and contacts in power engineering is presented in [12] based on [92]–[98]. According to this, copper is plated with a (1…2) μm nickel interlayer that serves as a barrier between tin (≥ 16 μm) and copper (substrate) (Tab. 6).

Tab. 6: Coating systems at $t = 0$ (measurements done at [1] IfWW, institute of materials science, Technische Universität Dresden and [2] fem, Research Institute for Precious Metals and Metal Chemistry, Schwäbisch Gmünd)

Coating process	System	Layer structure	Microstructure[1]	GDEOS[2] plot
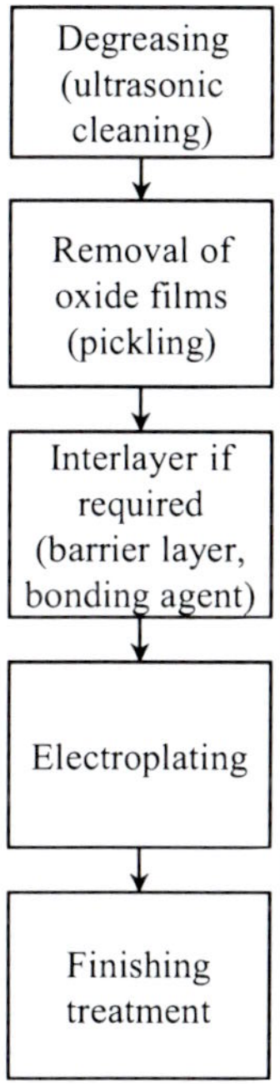	Cu-Ag	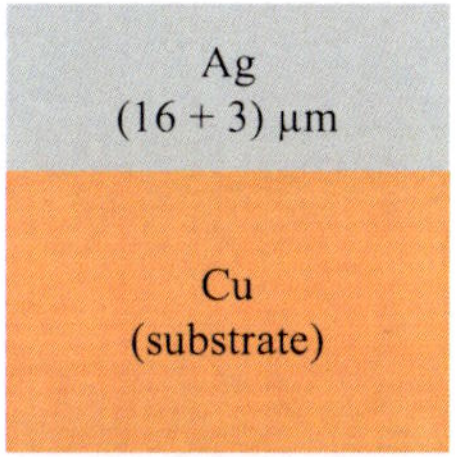	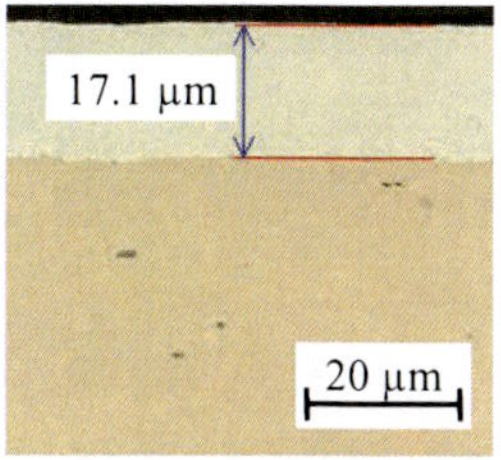	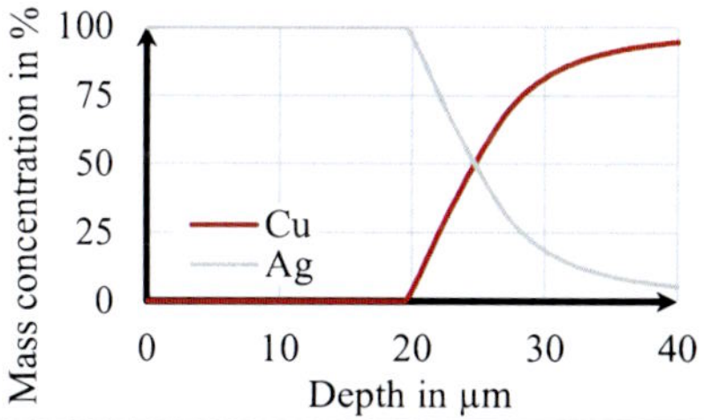
	Cu-Ni	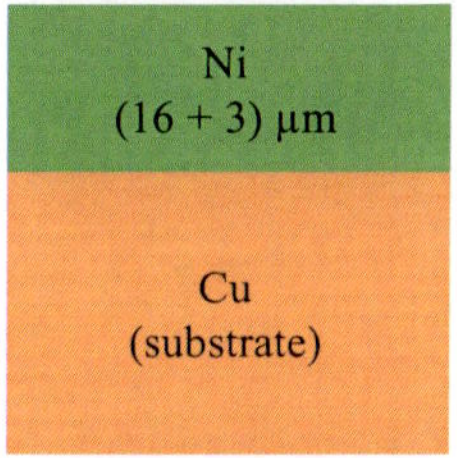	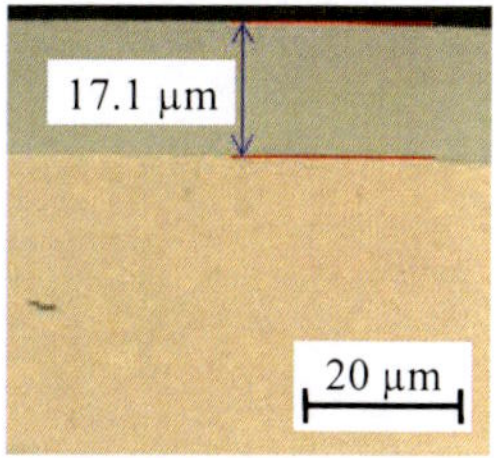	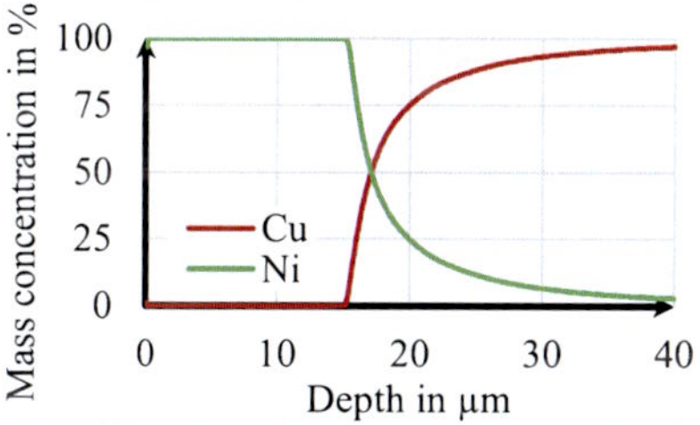
	Cu-Ni-Sn	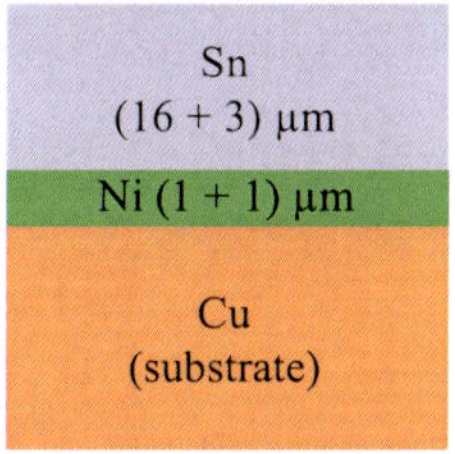	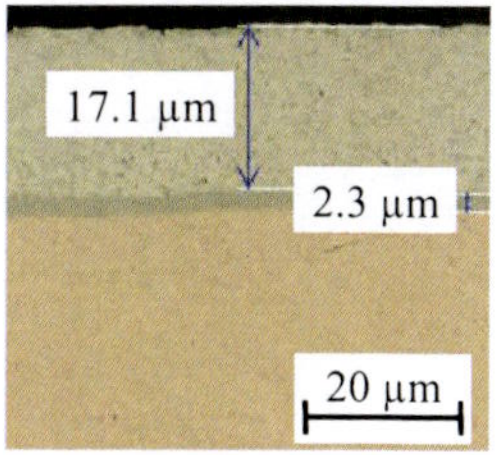	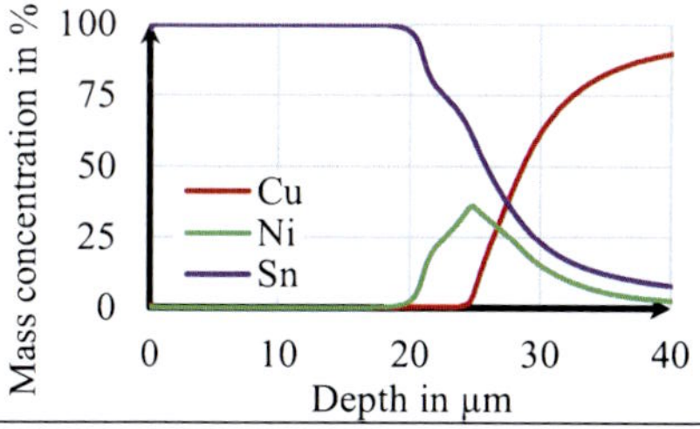

The chemical composition of the coatings was determined by a glow discharge optical emission spectroscopy (GDEOS) through surface sputtering and optical spectroscopic of the sputtered plasma (Tab. 6). All expected substrate and coating elements were identified. Beside this, some small amounts of carbon, phosphor and sulfur with less than 1 % of mass concentration were also identified. Typically these elements occur as dust. They not only can be found on surfaces but they could also be systemic depending on the measuring method. Obviously, no unwanted elements remain in the coating system from the coating process or somewhere else. The intersection point of the elements in the GEODS plot is a weak evidence for the layer's thickness. It consists a systemic error of (5…10) µm (Tab. 6). For Cu-Ni the intersection point of Cu and Ni differs between the two samples. This might indicate a less homogenous thickness along the surface. By visual inspections notches and / or pores (Appendix B, Fig. 75) were visible on the entire nickel plated surface.

For the top layer thickness + 3 µm tolerance was intended and for the sublayer a + 1 µm tolerance. The microstructure of Cu-Ag and Cu-Ni-Sn show that the determined layer thickness is within this range (the same thickness in Tab. 6 is random). A significant difference for the layer thickness of Ni in the Cu-Ni system of (10.4 … 28.6) µm was found in more detailed studies in Chapter 6.6 on the applied contact and material systems.

Physical Properties

A set of physical properties of the applied materials within this work is displayed in the table below. The literature is not distinct for all specified values. Some differences between several sources might be existing.

Tab. 7: Material properties with elastic modulus E, Poisson's ratio ν, density ϱ_{20}, softening point ϑ_s, melting point ϑ_m, coefficient of linear thermal expansion α_{th}, thermal conductivity λ, temperature coefficient electric resistance α_T

	Cu	Ag	Sn	Ni	Al	AlMgSi
E in GPa	115 [99]	79 [100]	47 [12]	216 [100]	70 [101]	70 [101]
ν	0.34 [99]	0.38 [102]	0.33 [103]	0.31 [100]	≈ 0.35 [103]	≈ 0.33 [103]
ϱ_{20} in g/cm^3	8.93 [99]	10.5 [7]	7.3 [7]	8.9 [100]	2.7 [101]	2.7 [101]
ϑ_s in °C	190 [7]	180 [7]	100 [7]	520 [7]	-	-
ϑ_m in °C	1083 [7]	960 [7]	232 [7]	1452 [7]	646…657 [43]	585…650 [43]
α_{th} in $10^6 \cdot K^{-1}$	16.7…18.3 [99] (20…207)°C	19.7 [100]	22 [104]	13.3 [100]	24.5 [43] (20…200)°C	
λ in W/(m·K)	≥ 380 [7]	418 [7]	64 [7]	70 [7]	210…220 [43]	200…220 [43]
α_T in 1/K	0.004 [7]	0.004 [7]	0.0045 [7]	0.005 [7]	0.0036 [105]	0.0042 [105]

Bulk Resistance

The bulk material's conductivity was determined by resistivity measurements on cylindrical rods and then compared with specifications according to standards. The measurement setup is as follows (Fig. 19):

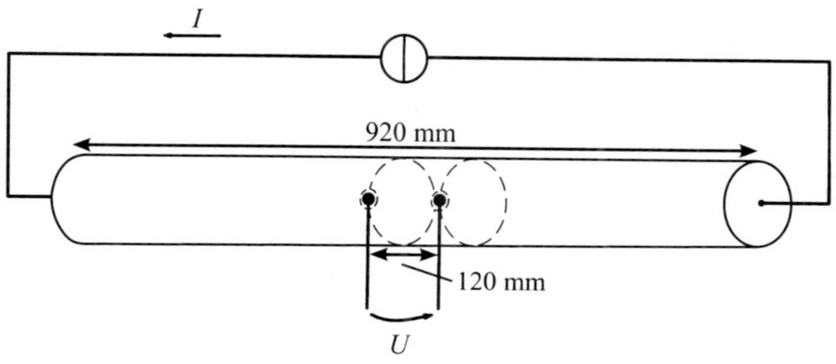

Fig. 19: Experimental setup bulk resistance measurement

The current feed-in was set up with a distance of about 900 mm in order to enable an undisturbed current distribution within the conductor. The resistivity (Tab. 8) was initially measured at ambient temperature and corrected to 20 °C with Eq. (30).

$$R_{20} = R(\vartheta) \cdot \left(1 + \alpha_\mathrm{T} \cdot (\vartheta - 20\,°\mathrm{C})\right)^{-1} \tag{30}$$

Tab. 8: Bulk resistivity and conductivity of Cu, Al, and AlMgSi at 20 °C, measurement done with microohmmeter LoRe (Werner Industrielle Elektronik, Appendix E)

$\rho_{20} = \frac{R(\vartheta)}{1+\alpha_\mathrm{T} \cdot (\vartheta - 20\,°\mathrm{C})} \cdot \frac{\pi \cdot r^2}{l}$	Cu			Al			AlMgSi		
	$\bar{\mu}$	$s\uparrow$	$s\downarrow$	$\bar{\mu}$	$s\uparrow$	$s\downarrow$	$\bar{\mu}$	$s\uparrow$	$s\downarrow$
ρ_{20} in μΩ·mm	17.4	17.3	17.4	28.1	28.0	28.3	31.6	31.6	31.8
$\kappa_{20} = 1 / \rho_{20}$ in $\mathrm{MS \cdot m^{-1}}$	57.6	57.9	57.4	35.6	35.8	35.4	31.6	31.7	31.5
κ_{20} in $\mathrm{MS \cdot m^{-1}}$ [31], [43], [101]	57	-	-	34...36	-	-	28..34	-	-

The conductivities for aluminum in [43], [101] are indications. The measured values of Cu and Al are within or close to the range given in [31], [43], [101] while the measured conductivity of AlMgSi is lower. The material data of the considered coatings were taken from the literature (Tab. 9). It can be seen in Tab. 9, that the resistivity of tin and nickel is distinctly larger than aluminum or copper. An exception to this is silver as it is almost equal to copper.

Tab. 9: Resistivity and conductivity of silver, tin and nickel at 20 °C

	Tin [12]	Silver [100]	Nickel [100]
ρ_{20} in μΩ·mm	116.3	15.9	68.4
$\kappa_{20} = 1 / \rho_{20}$ in $\mathrm{MS \cdot m^{-1}}$	8.6	62.9	14.6

6.2 Mechanical Stress

The connection system determines the acceptable mechanical load and the kind of force application on the electrical connections and contacts. As a result of the load application, elastic and plastic deformation occur at the contact area. The contact area deformation mode is determined by the contact force and might be not equal within the contact area. Moreover, the stress distribution may alter within the apparent contact area [12]. Obviously, it is useful to know about the mechanical stress level and how the stress distributes within the contact area. Significant different stress within the contact area may influence the possible rupture of films. Therefore, a stress distribution as function of the position might influence the long-term behavior of the considered contacts.

The mechanical stress can be distinguished according to its direction in normal stress perpendicular to a surface (this can be either compressive or tensile stress) and shear stresses tangential to a surface. Real stress often appear as multi-axial stress with individual stress components σ_{ij} and can be summarized in a stress tensor σ_{ij} (x,y,z):

$$\sigma_{ij}(x,y,z)=\begin{pmatrix}\sigma_{xx} & \sigma_{xy} & \sigma_{xz}\\ \sigma_{yx} & \sigma_{yy} & \sigma_{yz}\\ \sigma_{zx} & \sigma_{zy} & \sigma_{zz}\end{pmatrix} \tag{31}$$

According to the von Mises yield criterion, this can be described by an equivalent stress σ_e [103], [106]:

$$\sigma_e=\sqrt{\frac{1}{2}\left[\left(\sigma_{xx}-\sigma_{yy}\right)^2+\left(\sigma_{yy}-\sigma_{zz}\right)^2+\left(\sigma_{yy}-\sigma_{zz}\right)^2\right]+3\left(\sigma_{yy}{}^2+\sigma_{yy}{}^2+\sigma_{yy}{}^2\right)} \tag{32}$$

Usually deformation occurs at solids that where a mechanical load is applied. The deformation process is reversible if the deformation disappears after the load is removed (elastic). If the deformation remains after removing the force, it is irreversible (plastic). A full collapse of a component caused by deformation is called a fracture.

6.2.1 Contact Area Deformation

Considering a circular contact point with radius a and elastic deformation, the average pressure can be described by a root function in Eq. (22) with its maxima located in the center (Fig. 20). The pressure decreases from the center to the spot rim. If the applied force exceeds a certain level (approximately the yield strength $R_{p0.2}$), the materials response will be irreversible and plastic deformation occurs. The material would then be displaced until the mechanical stress is below the yield strength. Elastic deformation occurs until a certain level has been reached and is then transitioning into plastic deformation.

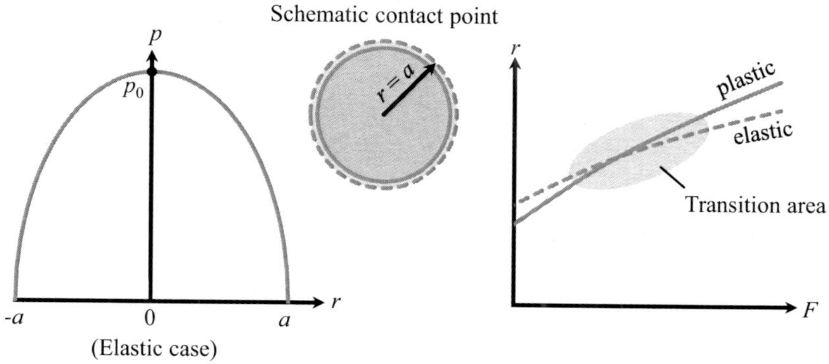

Fig. 20: Average pressure graph p dependent on the radius r (according to Eq. (23)) with maximum pressure p_0 and the radius r of the mechanically deformed area dependent on the force (according to Eqs. (22) and (24))

For the experimental studies in the chapters below crossed rods (diameter $d = 10$ mm) were used to establish circular contact point. The transition from elastic into plastic deformation of the considered contacts takes place at a few newton dependent on the material's properties. The analytical expressions for the radii of the elastic, Eq. (22), and plastic, Eq. (24), deformed area depends on a power law of second and third order. The plastic deformed area is rising faster when increasing the force F increased. The larger the force F, the lower the elastic ratio of the entire deformation. Beyond a significantly larger stress level that is than the yield strength $R_{p0.2}$, only plastic deformation occurs. If the applied force is far above the transition area, plastic deformation has the larger influence on the entire deformed area and becomes dominant. The elastic deformed area can be neglected in this case. The pressure in the plastic case cannot be easily determined by analytical approaches because of the material specific stiffness behavior. Numerical or experimental methods are convenient for this purpose.

6.2.2 Stress Distribution

In order to determine the stress distribution within the considered contact area, the finite element method (FEM) was used. The cylindrical geometry of the crossed rods was simplified to reduce computational time (Fig. 21). The model design process is outlined in Appendix C (Fig. 76). The numerical analysis software uses an implicit method to solve the model with 20-node hexahedron elements and an iterative solver (preconditioned conjugate gradient method).

The transitioning from elastic to plastic material behavior was realized by bilinear and multilinear isotropic material models (Fig. 22) and uses experimental determined true stress-strain curves from the literature. No meaningful stress-strain curves were found applicable to Al and AlMgSi in the literature that produce trustable results.

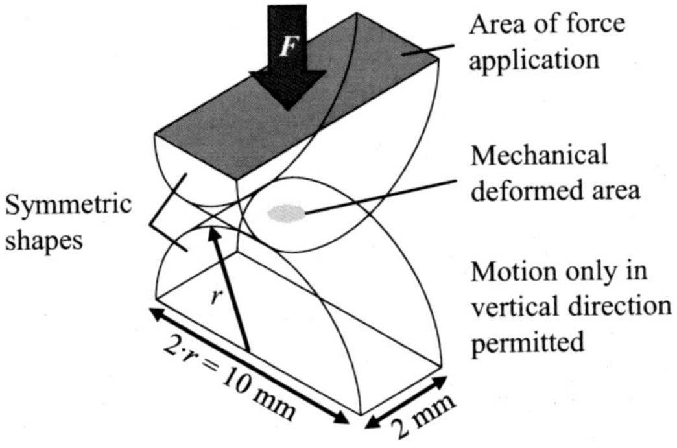

Fig. 21: Geometry model FEM

The model was tested with bilinear and multilinear material models with corresponding material data from [12], [107]–[110]. The experimentally determined stress-strain curves in [107] considerably spread for aluminum. The resulting radii of the plastic deformed area for Al and AlMgSi vary between experimentally, analytically and numerically determined results. The application of any experimental determined stress-strain curves from the literature was not possible in case of aluminum. However, the conclusion of the computational results might be qualitatively the same for aluminum and copper as long as only bulk material and no coating is present.

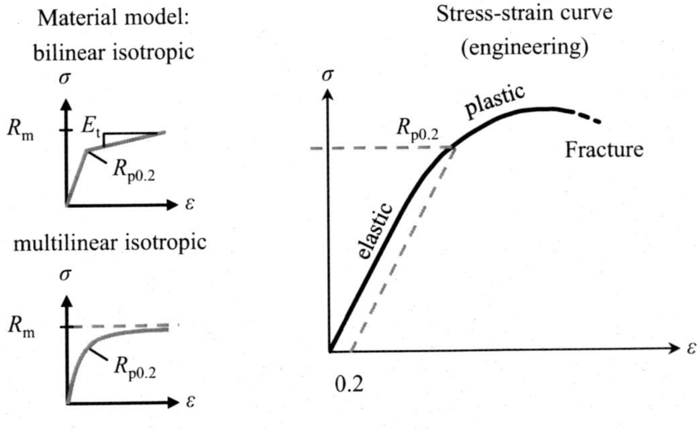

Fig. 22: Schematic of multilinear and bilinear material models with exemplary stress-strain curve for a real material with tangent modulus E_t, yield strength $R_{p0.2}$ and ultimate tensile strength R_m

The numerical analysis was performed with the mechanical load, applied in the experimental long-term studies ($F = 200$ N). The load was determined in preliminary experimental tests to establish reproducible test results and with reference to the test environments. The force is as high as contact forces in power engineering can be, but is above the force used for stationary connections in power engineering. However, the force is in the range of typical applications in electric power systems such as high power plug-in connections, disconnectors and circuit breakers for electrical loads according to [111].

For verification purposes cylindrical rods were used to obtain experimental results (test design see Chapter 6.3). It can be seen from the data in Tab. 10 that analytical and FEM results are close to visual determined radii. Considering acceptable deviations up to 10 % for the microscopy, the methods interfere in range.

Tab. 10: Radii of plastic deformed area determined by different approaches for crossed rods (Cu, d = 10 mm), material parameters were used from Tab. 7, Tab. 24, Appendix C

	Average radius plastic deformed area r_t ($r_{t\,min}$...$r_{t\,max}$) in µm		
Force F in N	$r_t = \sqrt{F/(\pi \cdot H_K)}$, H_K = (0.4...0.7) GPa [78]	FEM	Experimental
150	295 (261...345)	288	336 (309...360)
200	340 (302...399)	315	351 (322...382)
250	380 (337...446)	345	397 (375...415)

The qualitative results of the numerical analysis of the stress distribution (Fig. 23) reveals that the mechanical stress is higher in regions close to the contact area than those further apart. The results for Cu-Ni and Cu-Ag (Fig. 23) indicate higher stress prevails in the substrate compared to the coating. For Cu-Ni-Sn the highest stress levels appears in the substrate while the lowest level appear in the nickel sublayer. The coatings are directly at the interface area and possibly the deformation is therefore stronger. In addition, the coating thickness might have an impact on the mechanical behavior. Thin coatings can be dominated by the substrate's mechanical behavior. The substrate may dominate the mechanical behavior of the entire material systems.

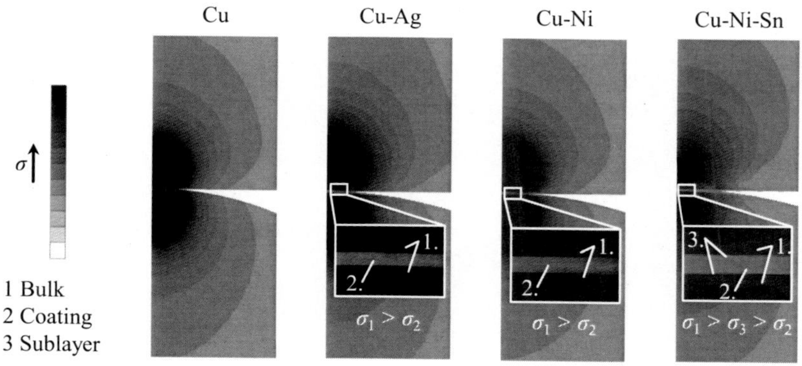

Fig. 23: Qualitative stress distribution according to von Mises criterion with bulk, coating and sublayer area and corresponding stress for a mechanical load F = 200 N, used material from Tab. 7, Tab. 24, Appendix C

For all cases outlined in Fig. 23 no significant difference of the stress distribution is observable in radial direction across and in vicinity of the contact area. Some visible but not significant differences may arise from singularities occurring as a result of the model solving process. The numerical analysis also reveals, that for thin layers and especially multiple-layers systems, the FEM computation possibly result in instable solutions.

Fig. 24 shows the stress for individual spots (dotted line) within the contact area computed by FEM. The grey circle indicate the area of a contact point. The elastic deformed share begins on the falling edge of the stress curve, whereas the plastic deformed area is mainly located on the plateau of the stress curve. As see in Fig. 24, a nearly equal stress distribution in the region of the contact point exists. The outcome is similar with previous work works [112]–[114]. The computed stress ranges between 210 and 240 N/mm^2 and is therefore slightly below the material models' ultimate tensile strength R_m. Because of high mechanical stress levels the material will be displaced or alternatively undergoes strain hardening.

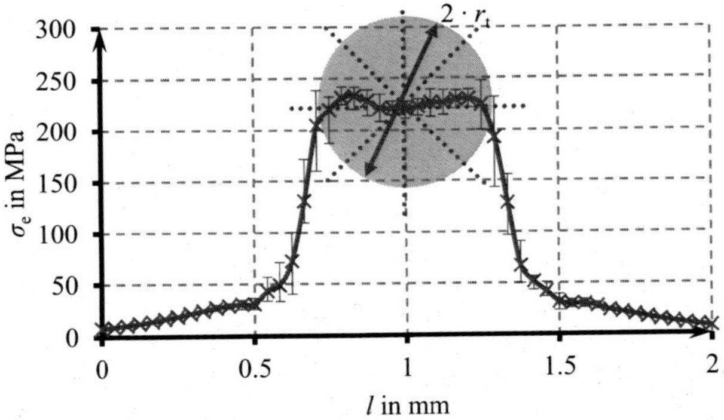

Fig. 24: Average equivalent stress distribution σ_e according to von Mises along horizontal, vertical and diagonal paths (dotted) across the mechanical deformed area computed by FEM and plan view of a Cu contact point for a mechanical load of 200 N (same as the load for the experimental long-term study), used material parameters see Appendix C

However, the FEM results might be useful for qualitative stress analysis. Considering deformation, precise material data would be more convenient. Moreover, the application of literature based material data may possibly result in inconsistent findings between experimental, analytical and FEM approaches.

6.3 Contact Resistance at Initial State

6.3.1 Experimental Design and Setup

Until today various methods have been developed and introduced in order to investigate electrical contacts, e.g. crossed rods [1], [3], [75], [82], [115]. For the intended tests crossed rods were chosen as samples. One of their advantages is their easy handling to measure the contact resistance by four-point probe. At very low forces, crossed rods cannot be used easily due to the very low mechanical stability. The necessary precision, required at very low forces, could be achieved in a laboratory environment but certainly not in any outdoor environment. A variety of preliminary studies were conducted to determine suitable dimension and loads. In line with the long-term study the dimension of the cylindrical rods suitable for any environment was set to the diameter $d = 10$ mm. The rods were manufactured by extrusion. A specific test design was constructed for this dimension (Fig. 25). The cylindrical rods were used to establish a circular contact point and contact each other only in vertical direction between the sockets. The contact force F was applied by a tension spring via the frame and adjusted through the elongation of the spring. In order to measure the contact resistance, the measuring current was fed into the rod's long end. The vibration-free voltage measurement was take at the currentless short end of the rods (Fig. 25).

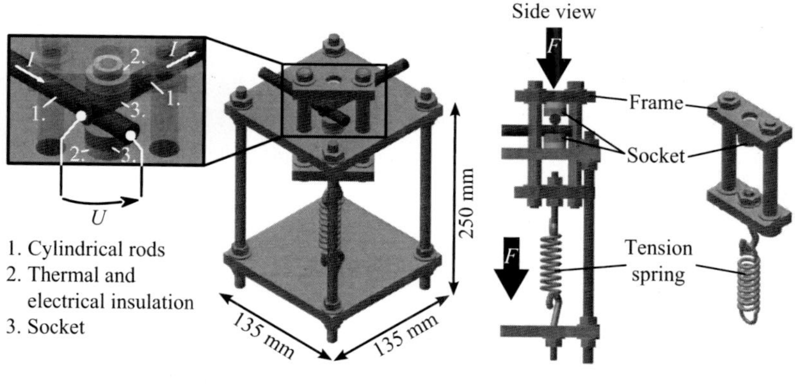

Fig. 25: Experimental setup crossed rods

Contact Force

The contact force determines the deformation mode in contact area. High contact forces are commonly applied in power engineering. Nevertheless for practical components, such as bolted busbar joints, the contact force per contact spot is not equal and significantly decreases from the inner to the outer side of the contact area [12].

The basic functional principle of the force also applies to interface areas of spheres or cylindrical rods. The deformation is pure elastic at low forces and becomes mainly plastic thereafter. The transition between elastic and plastic deformation for a given cylindrical copper contact with rods of diameter $d = 10$ mm begins at a very low force of a few N. The transition point is reached when the force has exceeded the materials yield strength. The center of the deformation would be plastic with the outer rim being elastic. The ratio between the elastic and plastic deformed area becomes smaller when increasing the force. It is therefore likely that for contacts with high contact force the dominating deformation mode of the contact area might be plastic.

It further appears that small contact radii are more sensitive due to changes in the contact area. Small contact radii are therefore intended for experimental investigations. They ultimately require an appropriate ratio between contact radii and contact force. As described above another test criterion is the mechanical stability of the experimental setup for outdoor conditions. In order to gain reproducible measurements, the contact force was determined in preliminary tests. Finally a load of 200 N was applied to ensure appropriate and stable contact resistances of crossed cylindrical rods with $d = 10$ mm.

6.3.2 Experimental Results at Initial State

The purpose of the experimental work conducted in this subchapter was to determine the initial state of the considered contacts. It was intended to create a database for comparison with the long-term results and to determine the influence of films on the considered contacts dependent on material and coating system at initial state. The resistance measurement was conducted at ambient temperature with temperature correction to 20 °C (according to Eq. (30)). Before taking the measurement, the rod's surface were cleaned with ethanol and a lint-free cloth as well as partially with a wire brush or an abrasive pad (Tab. 11). Surface cleaning before assembling is a standard proceeding for power connections in electric power systems to remove possible oxide films.

Tab. 11: Studied materials and cleaning procedure

Material	Cu	Al	AlMgSi	Cu		
				Cu-Ag	Cu-Ni-Sn	Cu-Ni
Cleaning process	Ethanol + abrasive pad*, grit size 200	Ethanol + wire brush	Ethanol + wire brush	Ethanol	Ethanol	Ethanol + abrasive pad*, grit size 1000

*(Scotch Brite™)

Certain measurements were taken at (150, 200, 250) N to test for the relationship between the force F, the contact area A and the contact resistance R ($R \sim F^{-m} \sim A^{-1}$). The exponent m is dependent on the mode of deformation. After the electrical measurements were taken the contacts were then opened and the interface area was microscoped to determine the plastic deformed contact area. The measurement and the evaluation were both performed based on the flow chart in Fig. 26.

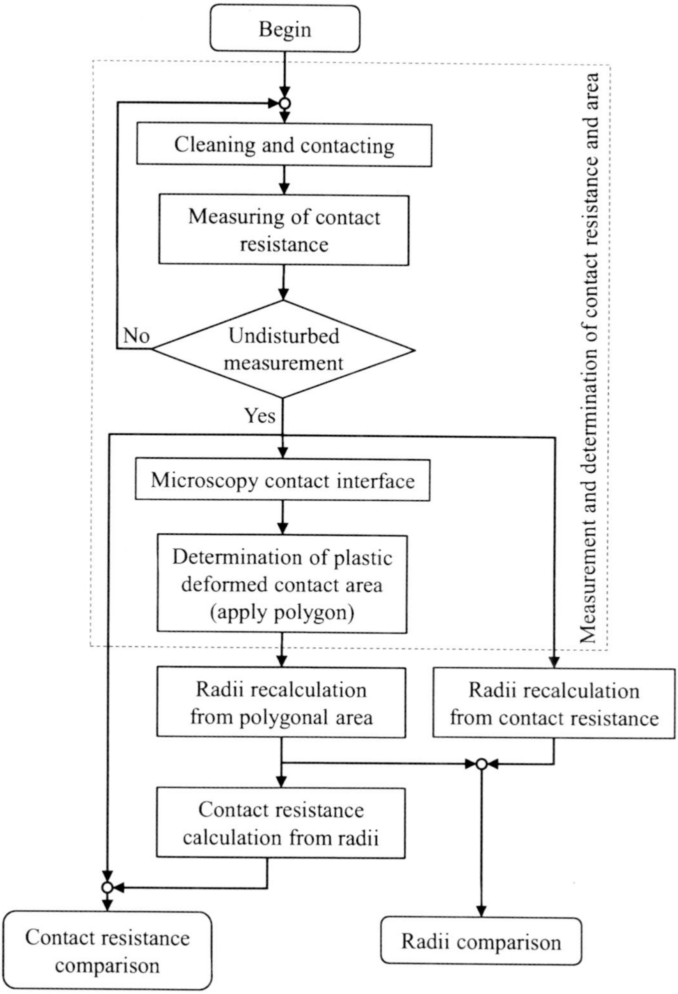

Fig. 26: Flow chart contact resistance measurement and microscopy at initial conditions

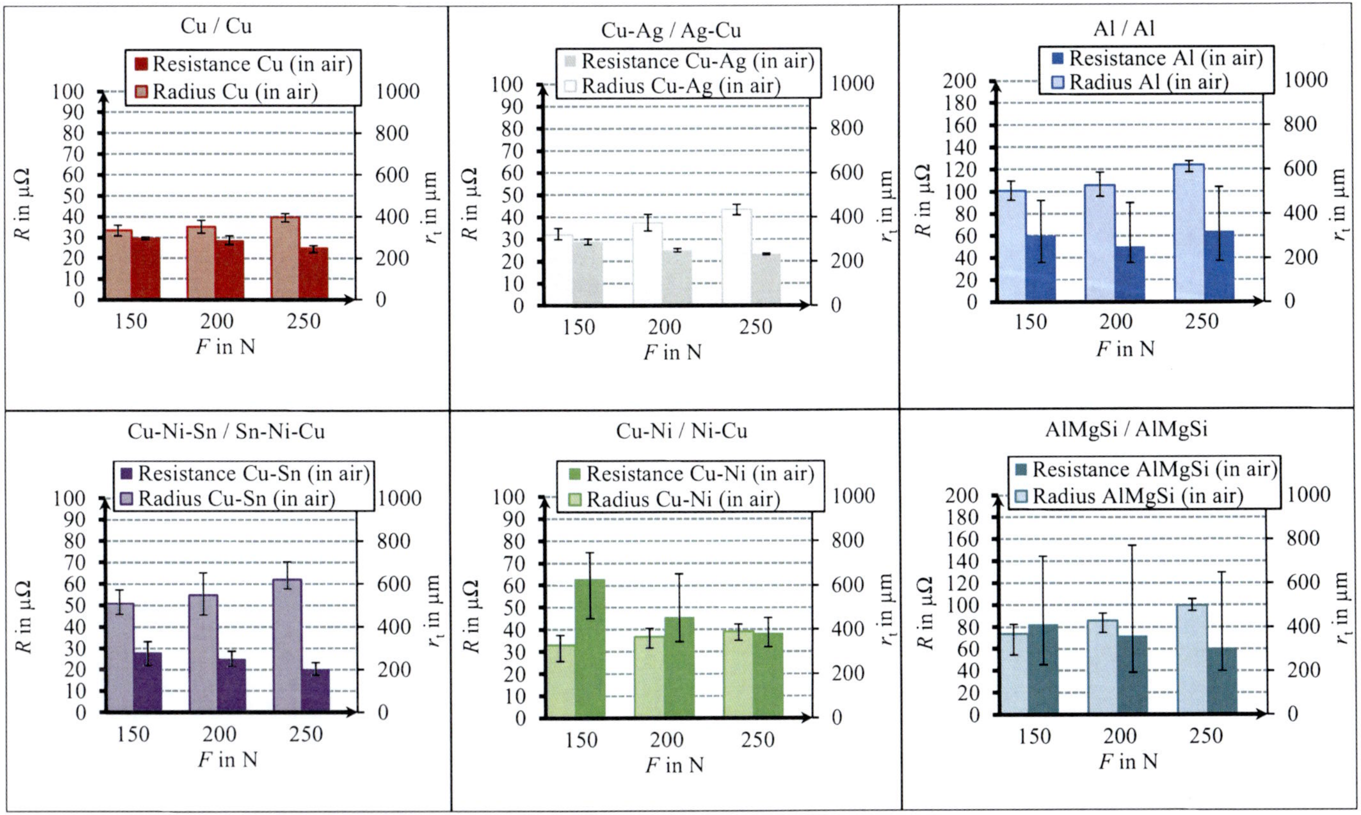

Fig. 27: Contact resistance and plastic deformed area dependent on force, / → contact between corresponding materials

Fig. 27 shows that the measured contact resistance of Cu, Cu-Ag, Cu-Ni-Sn and Cu-Ni decreases with increasing force. The reason for this are twofold: the larger contact areas by the mechanical deformation with $R \sim F^{-m}$ and secondly the rupture of possible films. The dependency of this material specific contact resistance behavior can be described by different power laws. For plated surfaces the exponent of such power laws is influenced by both, substrate and coating. The spreading of the measured contact resistance depends on the studied material and its native oxide film. After the cleaning process thin layers of oxide will form instantly on the metal surface. For Cu-Ni the oxide is quite hard and difficult to rupture. The presence of highly resistive films within the contact area causes quasi insulating areas and thus reduces the current-carrying contact area. The higher the force, the better the rupturing of the films.

The measured resistance of aluminum contacts (Al and AlMgSi) does not decrease with increasing force (Fig. 27). This is possibly caused by non-ruptured aluminum oxide films having a high resistivity. In Fig. 27 it can be seen, that the measured radii of the plastic deformed area increases with ascending force. Nevertheless, the increment of the plastic deformed area might be neutralized by the presence of high resistive films within the contact area.

Oxide films are harder than the native material and may result in an enhanced hardness of the contact surface. However, harder contact surfaces would result in smaller contact areas (Eq. (24)). Oxidized contact surfaces with thick films will probably result in smaller true contact areas and thereby in higher contact resistance.

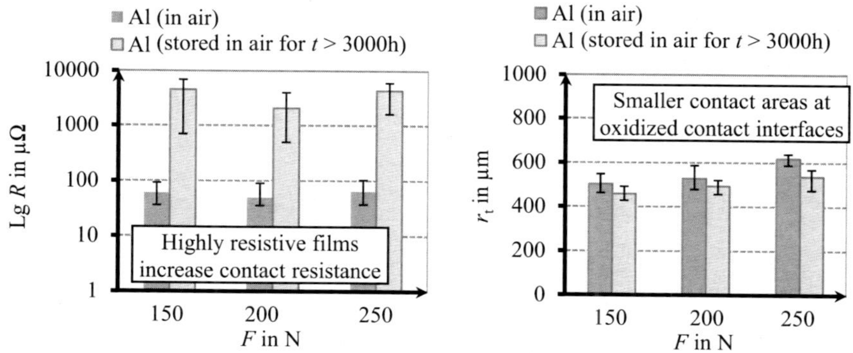

Fig. 28: Contact resistance R and radius plastic deformed area r_t of aluminum rods at oxidized conditions (stored in air for $t > 3000$ h at ambient temperature) and after cleaning with wire brush and ethanol, both measured in air

In order to test this assumption, several rods were oxidized at ambient temperature within a laboratory environment for a few thousand hours and contacted afterwards. The measured contact resistance is more than one magnitude higher compared to the measured resistance of freshly cleaned contacts (Fig. 28). In addition to that, the measured radii of the plastic deformed area is also smaller. This indicates that the contact resistance increase stems from high resistive films and a higher constriction. In return this implies that shortly after the contacting process has finished, the influence of films on electrical contacts at initial conditions occurs from the film's resistance at the contact zone. Secondly, it can be further concluded that smaller contact areas result in higher constriction and therefore higher contact resistance. The distinct difference between cleaned and oxidized surface can be also seen for crossed copper rods (Fig. 87, Appendix D).

In order to determine the electrical influence of films on the contact resistance at initial conditions, the actual measured values were compared with calculations based on the determined radii r_t of the plastic deformed area. From the microscopic images it can be seen, that multiple spots are present in the contact area of the crossed rods. The cluster radius α_c is supposed to be nearly equal to the radius of the plastic deformed area r_t. The spot number and the circular cluster area with $\alpha_c \approx r_t$ enable a simplification of Holms approach in Eq. (17) (see Appendix D). The contact points with cluster radius $\alpha_c \approx r_t$ were treated as individual spots and Eq. (7) with $a = \alpha_c = r_t$ (Eq. (33)) was used to perform the calculations in Fig. 29 for bare copper and aluminum contacts. For the calculation it is assumed that no films are present. The contact resistance with the bulk resistivity ρ_b and the measured radii r_t can be expressed as:

$$R = \frac{\rho_b}{2r_t} \qquad (33)$$

According to [1], applicable to tin and nickel with $\rho_{coating} >> \rho_{bulk}$ it is therefore acceptable to treat the coatings as films. The current field simplifies and Eq. (10) can be equivalently used [1]. The contact resistance in Eq. (34) consists of the corresponding resistivity ρ_b, ρ_{coa1}, ρ_{coa2}, the measured radii r_t and the thickness of the coatings s_1 and s_2. In case of pure nickel coating, the second term becomes zero ($s_2 = 0$).

$$R = \frac{\rho_b}{2r_t} + 2 \cdot \frac{\rho_{coa1} \cdot s_1}{\pi r_t^2} + 2 \cdot \frac{\rho_{coa2} \cdot s_2}{\pi r_t^2} \qquad (34)$$

Another approach was used for silver coated contacts by Eq. (35). Silver's resistivity is in the same range as copper. The contact resistance can therefore be expressed as a solution of the ellipsoid model with adapted intervals (Appendix D):

$$R = \frac{\rho_{coa}}{\pi r_t} \cdot \arctan\frac{s}{r_t} + \frac{\rho_b}{\pi r_t} \cdot \left(\frac{\pi}{2} - \arctan\frac{s}{r_t} \right) \qquad (35)$$

For silver coated contacts, the results of Eq. (35) are between 5 to 11 % closer to the measurements than Eq. (34).

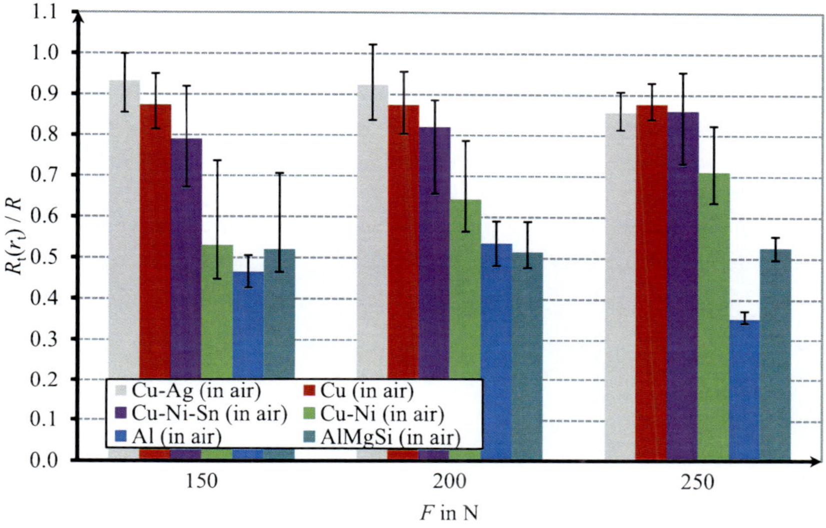

Fig. 29: Ratio between computed contact resistance $R_t(r_t)$ dependent on the plastic deformed area r_t and the mean of the measured contact resistance R; Cu, Al and AlMgSi determined with Eq. (33); Cu-Ni-Sn and Cu-Ni determined with Eq. (34); Cu-Ag determined with Eq. (35); data from Fig. 27 (r_t) and Tab. 8 (ρ_b for Cu, Al, AlMgSi) and Tab. 9 (ρ_{coa} for Ag, Ni, Sn) and Tab. 6 (thickness coating)

The results in Fig. 29 show that the normalized ratio $R_t(r_t) / R$ is close to one for Cu, Cu-Ag and Cu-Ni-Sn graphs. Compared to Cu-Ag and Cu, there is a stronger spreading of Cu-Ni-Sn which can be attributed to a larger variation in the determined radii r_t. The possible error caused by inaccurate radii determination is up to ± 10 % for all materials. Nevertheless no accordance is reached between the calculated and measured resistance for Cu-Ni, Al and AlMgSi. For Cu-Ni Fig. 29 shows that there has been a steady increase of the ratio $R_t(r_t) / R$ at increasing force. This might result from the increase of the current-carrying area because of a higher level of ruptured oxide films on the nickel surface. The results for Al and AlMgSi do not match with the calculations. Despite of the increased contact area (Fig. 28), the corresponding contact resistance spreads while being significantly below the theoretical resistance. The contact resistance deviation of up to - 15 % at oxidized aluminum cannot be explain by the smaller contact areas. The significant difference (Fig. 29) between calculated and measured contact resistance might arise from the high resistive oxide films that occur within the contact area. These not necessarily entirely covering films may work as insulation barrier and could either add an additional film resistance or they could also reduce the current-carrying area.

6.4 Studies in an Inert Gas Atmosphere

The development of films within the contact area of metallic conductors influences the long-term behavior of electrical contacts and connections. The possible ingrowth of films into the contact area can take place over long time durations. However, also the initial film formation might influence the long-term behavior for metals such as aluminum and aluminum alloys [79], [116]. Initial films could significantly increase the contact resistance for metals with brittle oxides and a strong attraction to oxygen. The following study tries to determine the quantitative influence of oxide films on aluminum in the initial phase after the contact has been established.

6.4.1 Experimental Design and Setup

In a standard atmosphere, oxide films would rapidly develop on a fresh cleaned aluminum surface according to Tab. 11. The film would be present in axial direction between the contact interfaces.

Atmospheres with reduced oxygen levels might significantly reduce the oxidation processes on such fresh cleaned aluminum surfaces. The focus in this study was pure aluminum Al. The experimental setup and the used geometry are the same as described in Chapter 6.3.1 (see also Fig. 25). The experiments were conducted in an inert gas atmosphere in a glove box. Inert gases are chemical passive and not very reactive compared with other gases like oxygen. In this study gaseous nitrogen N_2 was used. The experimental processes were cleaning, contact establishing, resistance measurement and the microscopy of the contact interface. Except the microscopy, all process steps were conducted in N_2 to reduce the oxygen influence (Fig. 30).

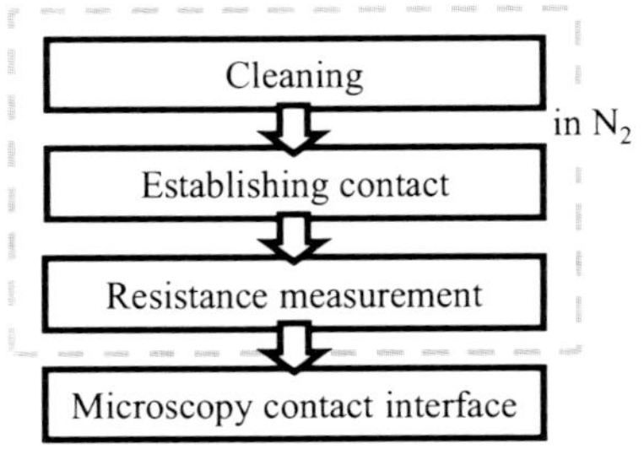

Fig. 30: Experimental proceeding and process steps within the inert gas atmosphere

The inert gas atmosphere was established within a glove box filled with gaseous nitrogen (Fig. 31). A constant positive pressure (greater than atmospheric pressure) was created by a vacuum pump to avoid external oxygen infiltration into the glove box. The concentration of unwanted gas components was kept small by a gas purification system.

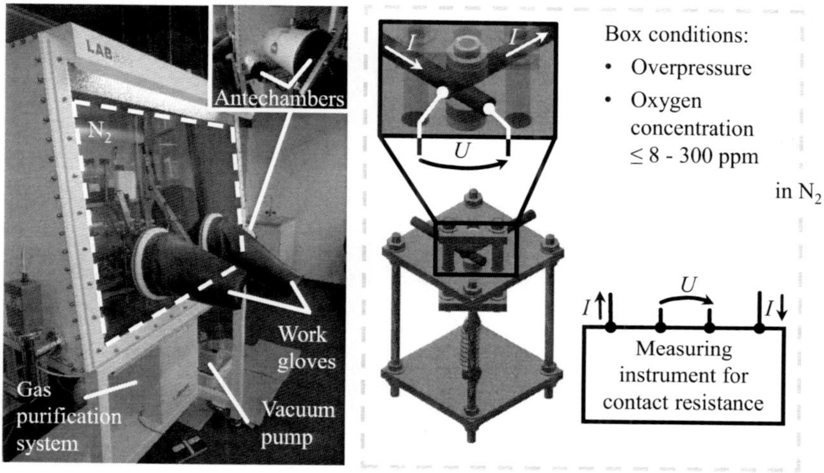

Fig. 31: Glovebox, experimental setup and devices, conditions inside the box, contact resistance was measured with a microohmmeter LoRe (Werner Industrielle Elektronik, Appendix E)

The oxygen concentration itself was permanently measured. The measured oxygen concentration during the tests was mainly between (8…100) ppm and occasionally up to 300 ppm. This amount is three orders lower than in ordinary air. The work glows were used for all experimental work inside the glove box. The samples and all required devices were transferred through the antechambers. The antechambers were evacuated and refiled with gaseous nitrogen several times during the transferring process. The experimental setup was verified with measurements on copper (Appendix F).

6.4.2 Experimental Process and Results

The experimental process was done at ambient temperature inside the glove box according to Fig. 30. The resistance measurement was temperature corrected to 20 °C according to Eq. (30). Certain measurements at 150 – 250 N were conducted for comparability with the results in Chapter 6.3. Before the measurement, the surfaces of the rods were cleaned according to Tab. 11 except the rods "Resistance Al 99.5 (336 h stored in air). These rods were mechanically processed by a lathe to remove the oxide films. Afterwards the rods were stored in a common laboratory atmosphere for ≈ 336 h. These rods were cleaned with ethanol only and no wire brush. The oxide film was therefore not ruptured. The strong influence of the oxide film on the contact resistance is shown in Fig. 32. As expected, distinctly lower resistances were measured for the rods with brushed surfaces. However, the experiments conducted in N_2 show even lower measured resistances as well as a lower spreading. The radius of the plastic deformed area was determined by microscopy. In Fig. 32, it is apparent that for less oxidized surfaces larger radii result. However, the difference between the measured radii is low and would not explain the differences between the measured resistances. All radii interfere in range.

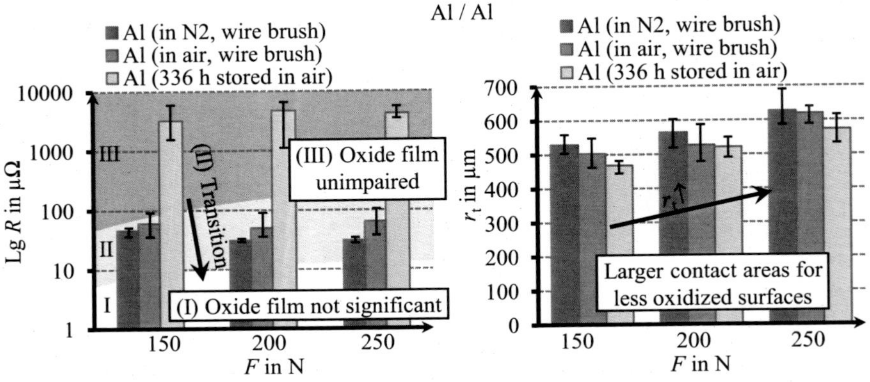

Fig. 32: Measured contact resistance R and radii r_t of Al contacts at different state, oxidized samples were stored in air at ambient temperature

As can be seen in the resistance graph in the figure above, three qualitative regions are characterized to describe the influence of the films within this regions. In the gray shaded area III the oxide film is thick and remains unimpaired. In case of aluminum the current can pass thinner oxide films with $s_f \leq (2\ldots5)$ nm only via tunneling. In area II the film is sporadically ruptured or even much thinner. Therefore, the influence of the film on the contact resistance reduces. As result a transition of the film dominated contact resistance occurs to a more constriction dominated contact resistance. In the last region, area I, the film is very thin, a monolayer or even not present. In this case, the remaining tunnel resistance is very low (according to [3], [66] , Eqs. (12) – (15)). The transition of area I – III is smooth and may differ for miscellaneous materials and further contact parameters.

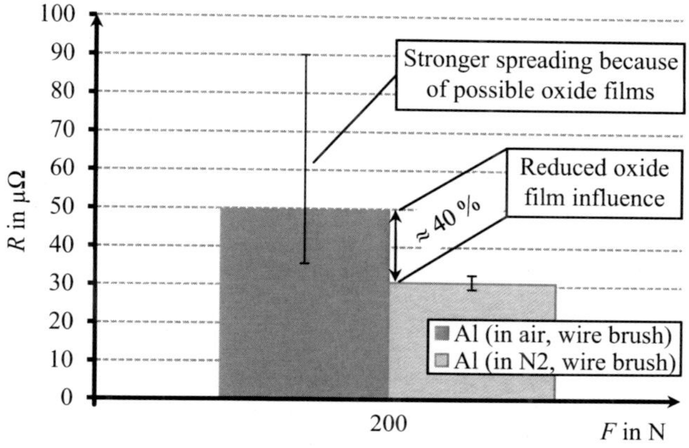

Fig. 33: Measured resistance of Al contacts with equal cleaning in air and gaseous nitrogen

Fig. 33 reveals that there is a significant difference of up to 40 % between the contacts measured in air and in N_2. Both variants have been equal cleaned. The concentration of oxygen within the glovebox was < 0.03 %. Physisorption or chemisorption processes of oxygen are reduced because of the low concentration of oxygen. In Fig. 33 the spreading of the measurements in air show the influence of an oxide film with a variable thickness. These films significantly increase the film resistance. If the films are thick enough to prevent tunneling, they might reduce the current-carrying contact area and increase the constriction resistance. Eventually the films might influence the contact resistance in both ways.

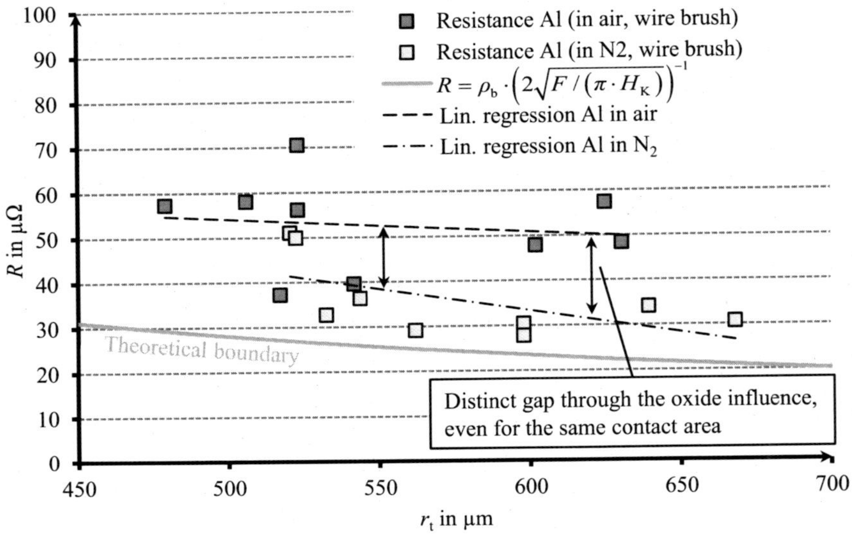

Fig. 34: Contact resistance dependent on the radius of the plastic deformed area r_t plotted for individual measurements and regressions as well as contact resistance R determined according to Eqs. (24) and (33), $H_K = 400$ N/mm² [78], $\rho_b = 28.1$ µΩmm and variable force F

In Fig. 34 the observed gap between the measurements done in air and N_2 possibly results from the influence of the films which are less present in N_2. This fact is obvious when comparing measurements with the same area but different resistance. The theoretical boundary is an approximation of the contact resistance with no oxide films. Indeed the exact path of the approximation may differ. The closer the measurements to the theoretical boundary, the less is the influence of the oxide films. Some deviations might come from disturbances in the measurement proceeding. Further measurements with AlMgSi and Cu-Ni were also conducted but were evaluated because of too little sample size. During the experimental work, it was observed that the oxygen concentration increased if ethanol was used: The concentration increase was within the above addressed limits (Fig. 31). The effect was some more distinctive if ethanol was in contact with the work gloves. Possibly the ethanol evaporation and the contact with the rubber gloves may result in new free oxygen molecules.

6.5 Impact of Film Development during the Formation Phase

Within the formation phase (see also Fig. 56), stationary electrical contacts and connections are exposed to its operating temperatures for the first time. The operating temperatures are determined by the ambient temperatures and joule heating inside the metallic conductors. Usually these temperatures are high enough to induce a heat treatment. Temperature dependent physical processes in solids will be activated during the heat treatment [11]. Such physical processes also depend on time and material.

From prior works it is known, that the measured contact resistances might be changing within the formation phase [12], [13], [83], [117]. Considering the case of copper, the measured contact resistance decreases in the formation phase. It is assumed that larger current-carrying contact areas develop. These larger areas are induced by physical processes in solids as a result of the heat treatment. In turn, larger current-carrying contact areas will reduce the contact resistance. The decrease of the contact or connection resistance in the formation phase was observed in several long-term studies [12], [79], [83]. The effect occurred in aging tests in air with current and Joule heating as well as in aging tests in oven in air (currentless).

Parallel to the formation phase, a significant film growth occurs on the fresh cleaned surface. At steady conditions, the film growth is strongest in the initial period after cleaning and begin of thermal operation. It is supposed that interactions exist between film growth and physical processes in solid-states dependent on time, material, and temperature. The influence of these interactions on stationary electrical contacts and connections are not fully understood for the formation phase.

6.5.1 Experimental Design and Setup

The considered material for this study was Cu-ETP. The same geometry of crossed rods was used with the same dimensions ($d = 10$ mm) as described in Chapter 6.3.1 (Fig. 25). The cleaning was done with an abrasive pad (grit size 200) and ethanol according to Tab. 11. Three test assemblies were stored in an oven for 1000 h at 200 °C (Fig. 35). The film development at 200 °C is faster according to Eq. (27). The considered degradation mechanism should not significantly change in this temperature range according to previous research [82], [83]. As observed by Rönnquist [26], it is expected that mainly copper(I) oxide Cu_2O should develop within this temperature range.

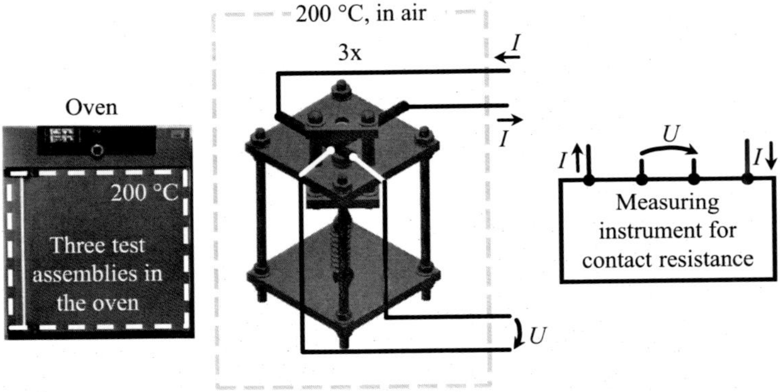

Fig. 35: Oven and experimental setup as well as test assemblies and devices, the contact resistance was measured with a microohmmeter LoRe (Werner Industrielle Elektronik, Appendix E)

6.5.2 Surface Oxidation and Temperature Corrected Contact Resistance

The contact resistance measurements were conducted at ambient temperature and during operation at 200 °C. The thermal operation was interrupted for measurement at ambient temperature. The first measurement was done at ambient temperature only. All measurements were temperature corrected to 20 °C with Eq. (30). In [1] Holm used $0.67 \cdot \alpha_T$ instead of α_T to determine the contact resistance with reference to a temperature. $0.67 \cdot \alpha_T$ does apply for non-steady state of contacts with Joule heating and high temperature difference between conductor and contact. This is not the case for the considered tests in oven or generally the resistance measurements of inactive steady circuits at ambient temperatures. Here $1 \cdot \alpha_T$ was used in preference of $0.67 \cdot \alpha_T$.

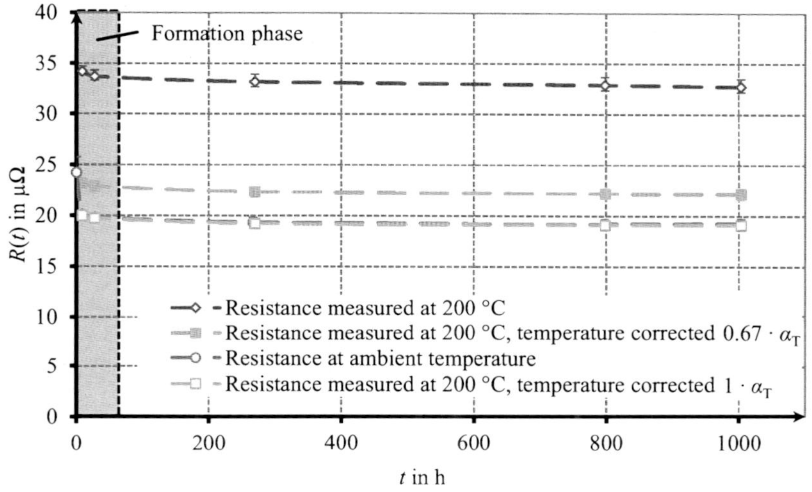

Fig. 36: Contact resistance R of crossed copper rods measured at 200 °C and ambient temperatures, both dependent on time t and temperature (200 °C), temperature reference according to Eq. (30)

The impact of the temperature correction can be seen in the difference in Fig. 36 between the graphs measured at 200 °C and at ambient temperature. Both are corrected to 20 °C. The measurement at ambient temperature is nearly equal to the graph, measured at 200 °C and corrected to 20 °C (plotted in Fig. 36, partially covered). If possible films in the contact area would dominate the current flow, the temperature coefficient would be inverse. Using - α_T with the experimental data from Fig. 36, the determined resistances would result in false resistances (not plotted). The resistance is therefore dominated by the bulk material. In Fig. 36 there is a clear trend of decreasing resistance within the first 100 hours. Afterwards, the resistance remains more or less stable.

The surface was inspected regularly together with the resistance measurement. At $t = 0$, the surface was bright and typical copper-colored (Fig. 37). Shortly afterwards the surface color was changing to reddish, partially violet and subsequently brownish. Later, the surface color became darker and eventually black. Since 750 h operation time, small areas of the now black cover layer spalled from the surface. Below the spalled surface areas, reddish and violet areas were observed. According to [26] (see also Fig. 14), only Cu_2O will develop in standard atmosphere at 200 °C. However, the literature is not clear but even when CuO appear it might not become dominant compared with Cu_2O.

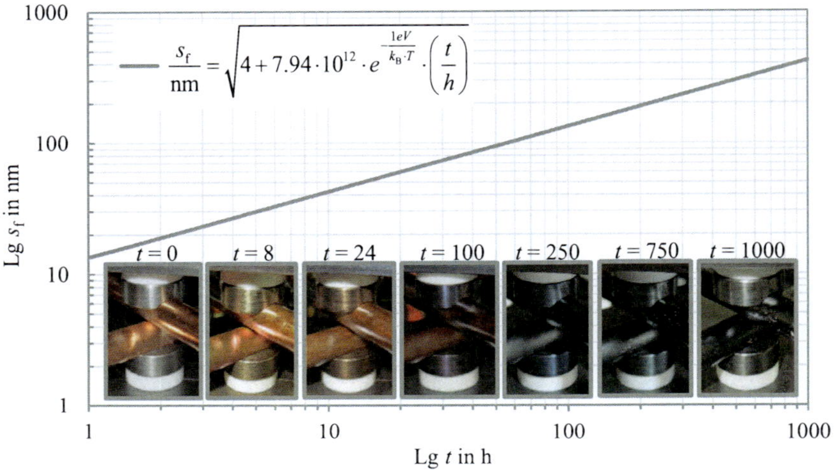

Fig. 37: Film thickness s_f of Cu_2O according to Eq. (27) and surface conditions of crossed copper rods, both dependent on time and temperature (200 °C); the activation energy is 1 eV or 96 kJ · mol^{-1}, k_B is the Boltzmann constant, T is the temperature in K

With Eq. (27) [1], the film thickness s in nm was determined in Fig. 37. A constant is added to the equation for the faster initial film growth and for $s(t = 0) = 2$ nm. The growth rate decreases over time as long as the film remains intact. If the film would be penetrated by mechanical stress, the copper surface could be exposed to atmospheric air again and the growth rate would increase again. The results of the optical inspection (Fig. 37) indicate a strong film growth which might be up to 0.5 μm according to Eq. (27).

6.5.3 Influence of Simultaneous Film Growth and Material Yielding on the Contact Resistance

In Fig. 36 there is a clear trend of stable contact resistances after the formation phase. Although it can be seen from Fig. 37 that a strong film was grown on the surface. Fig. 38 provides the determined radii of the current-carrying contact area dependent on time. Conversely to the resistance, the determined radii increase in the formation phase. After 1000 h testing time the heat treating was finished. Subsequently, at ambient temperature the contacts were opened and microscopic analysis were conducted according to Fig. 26. In Fig. 38 the two microscopic images I and II represent the same contact. Close inspection of the microscopic images reveal different results. The entire plastic deformed area (dashed circle, Fig. 38, II) consists of areas without noticeable films in radial direction towards the contact area center. In the center a more reddish area can be identified (continuous circle, Fig. 38, II). More detailed microscopic inspection showed that parts of the entire plastic deformed area were possibly covered by a thick film before being deformed (Appendix G, Fig. 91). The thermally activated material yielding process enlarged the plastic deformed area and decreased the contact resistance. This process proceeded until the film on the surface could not be penetrated anymore and became highly resistive. At this point the increase of area does not contribute to the resistance decrease and the contact resistance remains stable. This effect might not occur in N_2 when far less oxygen is present and the film growth would be reduced. In this case, the process would continue until the mechanical stress is too low to resume the material yielding.

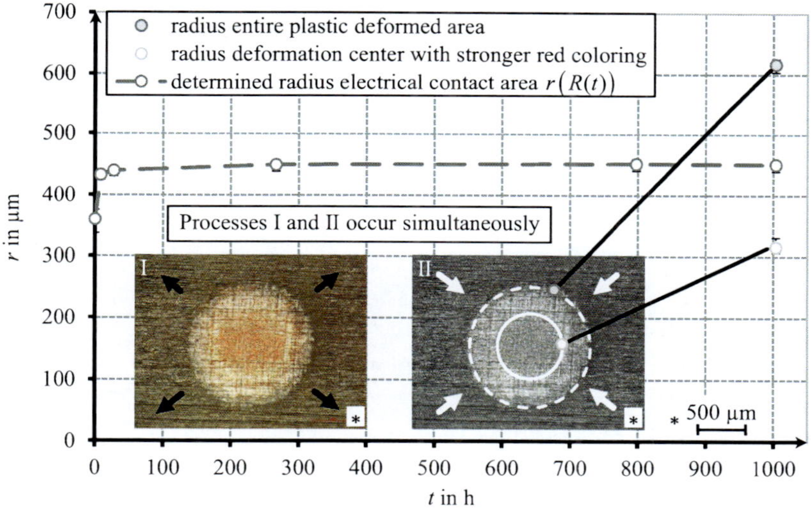

Fig. 38: Radius dependent on time and microscope image I and II at $t = 1000$ h; $r(R(t))$ was determined with Eq. (5), $\mu \rightarrow \infty$, process I: Thermally activated material yielding; process II: Slowly radial film growth towards the contact area center; process I dominates process II

After the formation phase, the further radial film growth is slowly and probably depends on the diffusion of oxygen into the oxide free contact area. The thicker the film the lower the film growth. Obviously both processes (I and II, Fig. 38) occur simultaneously. The material yielding is increasing the contact area until the film growth prevented the increase of oxide free contact area. Overall, these results thus indicate an equilibrium between material yielding and the oxide film growth process. Even the material yielding process continues, the true contact area did not longer increased. The contact resistance remain constant.

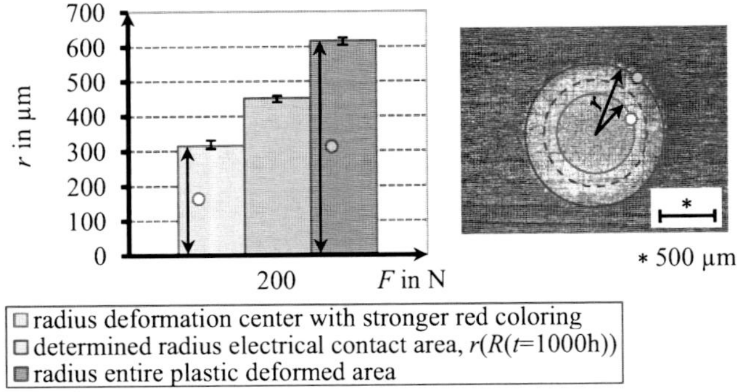

Fig. 39: Radii dependent on force; microscopic image with measured radii (microscopy) and radius $r(R(t = 1000\text{ h}))$ from resistance measurement (dashed line) determined by Eq. (5), $\mu \to \infty$

Fig. 39 compares the determined and measured radii based on the same data as in Fig. 38. The inner radius at the deformation center with stronger red coloring is approximately in the same range such as the experimental determined radii in Chapter 6.3. From the graph above, it can be seen that the radius of the determined electrical contact area (dashed, Fig. 39) is between the radii of the deformation center and the entire plastic deformed area. The results suggest, that the current flow in the here considered contact area is possible across metallic spots in the deformation center. In radial direction from the deformation center and increasing r (microscopic image Fig. 39), the electric current transport will possibly occur across thin tunnel films. For thicker films no current flow will occur. Overall these results show that even in the formation phase the film growth was restricting the resistance decrease by thermally activated material yielding processes.

6.6 Long-Term Behavior of Stationary Electrical Contacts with Reference to Film Development

In this chapter experimental long-term studies were carried out to determine the influence of films on the long-term behavior of stationary electrical contacts. With respect to the different degradation mechanisms the needed test geometry should ensure a susceptibility to films only. Other degradation mechanisms should be excluded somehow or significantly suppressed to ascertain the true influence of the films. The influence of different environments is considered to create a film promoting atmosphere based on non-artificial environmental conditions. The operation of the experimental study was carried out with respect to the permissible temperatures in accordance with IEC 62271-1: 2007 + A1: 2011.

6.6.1 Experimental Design and Setup

According to Chapter 6.1, the used materials for the crossed rods were Cu, Al and AlMgSi as bulk materials. Several copper rods, plated with silver (Cu-Ag), tin (Cu-Ni-Sn) or nickel (Cu-Ni) were used.

Geometry and Dimension

The dimensions of the rods are the same as described in Chapter 6.3.1 with diameter $d = 10$ mm. For each material combination, a number of contact pairs was serially connected in a test assembly (Fig. 40).

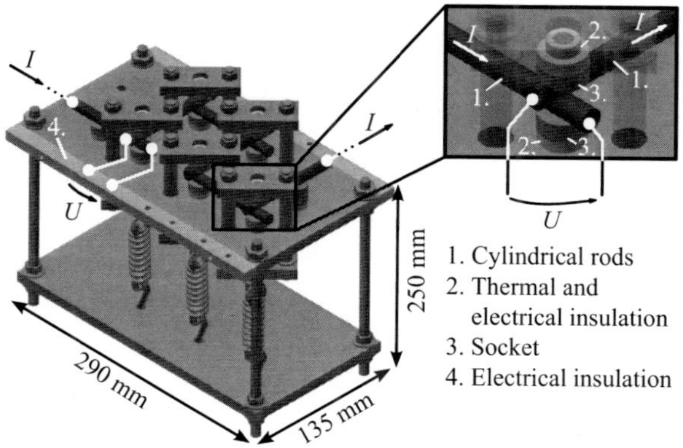

Fig. 40: Test assembly with five contacts connected in series and detailed view of an individual contact pair of cylindrical rods

The test assembly is based on the same functional principle as that one described in Chapter 6.3.1. The cylindrical design of the rods facilitates a good interaction between the environment and the contact area to form oxide or corrosion films. Likewise it was further necessary to pay attention to the mechanical stability. Each two rods contact each other in one point per contact. The superior rod is hold whereas the lower rod is laid in a milled stainless steel socket. A further MACOR® socket rests on every stainless steel socket as thermal and electrical insulation. The macor sockets on the other hand are deployed on the ground plate (lower) or being hold by the frame (superior). During operation the applied current commutated from one rod to the other one over the contact interface. The projecting ends of the rods are used as voltage taps. Wires were fixed at the projecting ends to enable the voltage measurement free of mechanical stress. During contact resistance measurement the circuit was fed with a measuring current and the resulting potential per contact was determined at the voltage taps.

For copper rods, five contacts were connected in series. In Chapter 6.3 a stronger scattering for aluminum was observed. Therefore, for Al and AlMgSi, ten contacts each were connected in series (experimental setup laboratory 90 °C, Appendix H, Fig. 92) in a larger test assembly.

The designed test geometry ensures a possible susceptibility to films as result of chemical reactions only. As well good interaction with the surrounding environment is assured to make the environmental influence on the contact resistance, if ascertainable, visual.

Thermal Load

The thermal load for the long-term tests was determined in accordance with the IEC 62271-1:2007+A1:2011 at 90 °C (highest temperature for bare cooper connections). It is known, that dependent on the temperature the film development is more rapid. To promote an accelerated film growth a temperature of 200 °C (concordant with Chapter 6.5.1) is selected for a limited number of test samples in a laboratory environment. Tin was investigated at 90 °C only because of its low melting temperature. According to previous investigations and the findings of the literature, the physical processes of the film development should not significantly change within the considered temperature range [1], [4], [12], [50], [82], [83].

Tab. 12: Investigated material combinations and operating temperature with number of samples

	Cu	Cu-Ni-Sn	Cu-Ag	Cu-Ni	Al	AlMgSi
90 °C	5	5	5	5	10	10
200 °C	5	-	5	5	10	10

During operation the temperature was controlled and therefore measured via thermocouples. The top of the thermocouples were punched in the bulk material of a rod at a position in (5…10) mm distance to the contact interface.

Contact Force

The force was adjusted by a tension spring. In almost the same manner as described in Chapter 6.3 the spring is elastic moveable in vertical direction (Fig. 41) and applies the force F on the contact dependent on the spring elongation. The selected springs consist of stainless spring steel and have a low spring constant. The spring constant was determined in preliminary studies and is ≈ (11…17) N/mm. The used spring constants were randomly controlled. No significant deviation was observed.

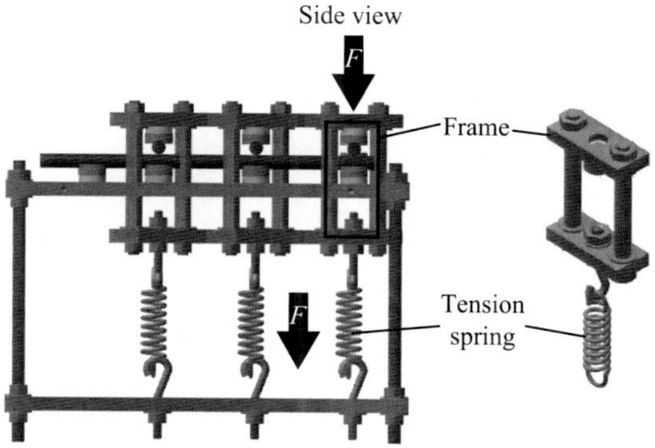

Fig. 41: Test assembly and individual frame, the tension spring pulls the frame → force F is applied on the contact

The results obtained from the deformation analysis of a copper contact at 200 N computed by FEM suggest a deformation in axial direction of less than 15 µm. However, this would mean a force adjustment of (0.17…0.26) N of the used tension spring in the experimental study. This outcome indicates a negligible influence of the deformation distance.

Cleaning

All contact pairs were cleaned before being in contact according to Tab. 11. The resistance was controlled after contact establishing by measurement. For aluminum (Al and AlMgSi) only resistances ≤ 60 µΩ were accepted to ensure nearly similar thermal conditions at the contacts during operation. Contacts with higher resistance were opened and cleaned again.

Environment

The experiments were carried out in different environments in Dresden presented in Tab. 13, namely: laboratory (indoor), botanical garden (tropical rainforest), and outdoor. It was intended to determine whether the different environmental conditions would have an impact on the performance of the electrical contacts. In the laboratory two sets were constructed and operated at 90 °C (Appendix H, Fig. 92) and 200 °C (Chapter 6.6.4). The samples in the Botanical Garden of the Technische Universität Dresden (TUD) were exposed to a tropical rainforest environment (Appendix H, Fig. 93 and Fig. 94). The outdoor set is constructed at an outdoor test site nearby a medium voltage switchgear station inside the municipal area of Dresden (Appendix H, Fig. 93 and Fig. 95). Climatic data were gathered from sensors close to the test set positions respectively. The laboratory environment remains more or less constant with a mean temperature of 24 °C and a relative humidity of 33 %. The mean temperature and the mean relative humidity of the botanical garden are 26 °C and 58 % (see also Fig. 58, Fig. 59). The measured data of the outdoor environment present the largest spread. The measured mean temperature is at 12 °C and the mean relative humidity is at 75 %. Strong extreme conditions with temperatures of minus 13 °C occur only rarely.

Tab. 13: Considered environments for long-term studies with mean (min...max) temperature and relative humidity

Environment	Botanical Garden	Outdoor	Laboratory	
Thermal load in °C	90	90	90	200
Ambient temperature in °C	26 (15...45)	12 (- 13...35)	24 (20...32)	
Relative humidity in %	58 (23...85)	75 (23...100)	33 (20...42)	

6.6.2 Impact of other Degradation Mechanisms

Force Reduction

For stationary contacts force reduction is a mayor degradation mechanism. Because of the spring characteristic of the contact assembly, force reduction was suppressed. The low spring constant would result in a very small force response only, when changing the elongation. Therefore, the force is nearly constant during the testing time.

Intermetallic Compounds

It is known that different firmly bonded metals tend to establish intermetallic compounds. Intermetallic compounds are often brittle and highly resistive. The electric contact resistance can be significantly increased. In case of copper and tin a (1...2) µm nickel sublayer was plated between Cu and Sn as diffusion barrier. However, Ni and Sn might also tend to develop intermetallic compounds. For the other considered material combinations, the presence of significant intermetallic compounds was not expected.

Electromigration

The degradation mechanism Electromigration is associated with the application of current load. This is reasoned as follows: Electromigration occurs mainly at DC load and considerably less at AC together with high current density. Nevertheless some observations of AC caused electromigration degradation had been made [23] whereas for other investigations [118] no aging due AC induced electromigration was observed. In this study AC was applied. The possible influence of AC induced electro migration was determined according to the findings in [16], [18]–[23]. No possible significant influence on the long-term behavior of the considered aluminum contacts was determined.

Sliding Wear and Fretting

AC or high degrees of freedom in motion can result in small motions of the current-carrying parts. Stationary contacts are characterized by high force and thus higher mechanical stability. The here considered experimental study is focused on stationary contacts and were carried out in a way that a movement of the current-carrying parts is limited to a very low level.

6.6.3 Long-Term Behavior of Crossed Rod Contacts Dependent on the Environment at 90 °C

The experimental tests were operated for 12000 h in the considered environments (Tab. 13) with the experimental setup pictured in Fig. 92 and Fig. 93 (Appendix H). The experimental tests were switched off for regularly measurement of the contact resistance. In Fig. 42 and Fig. 43, the measured contact resistances are temperature corrected according to Eq. (30). Fig. 42 shows that no significant resistance increase was observed for any of the considered environments. Cu-Ni-Sn presented the lowest measured resistance dependent on time and Al and AlMgSi either the highest. In Fig. 43 the measured resistances in different environments are plotted for each material system. The figure reveals that no significant difference of the measured resistance dependent on the environment was observed for Cu, Cu-Ag and Cu-Ni-Sn. Cu-Ni, Al and AlMgSi scatter stronger because of higher difference in initial resistance. The range remains more or less constant beyond the formation phase in the first 1000 h.

As can be seen in Fig. 42 and Fig. 43, in the formation phase the measured resistance drops for all selected materials systems even for aluminum (Al and AlMgSi). Finally all contacts eventually have stable measured resistance after 12000 h (Tab. 14). The measurements indicate that the results of the different environments may be similar. In [79] it is assumed that the rise of the connection resistance of aluminum busbars joints in the formation phase is associated with film development by chemical reactions. In this study no increase of the contact resistance was observed in the formation phase, in fact, the contact resistance decreases. This might come from the stable contact force which is not the case for each technical electrical connection.

Moreover, it can be seen, that the tin coating on Cu-Ni-Sn is almost entirely displaced (Fig. 46) from the contact area. Only some fragments remain on the nickel sublayer. In the non-contact area possible intermetallics develop with no impact on the contact resistance.

Tab. 14: Mean measured resistance $R(t)$ at $t = 12000$ h (≈ 1.4 a) at 90 °C and initial state $t = 0$; measurements are temperature corrected to 20 °C according to Eq. (30)

Environment	Laboratory		Botanical Garden		Outdoor	
Thermal load: 90 °C	$R(t)$ in µΩ					
	$t = 0$	$t = 1.4$ a	$t = 0$	$t = 1.4$ a	$t = 0$	$t = 1.4$ a
Cu / Cu	23.1	20.0	25.9	20.9	21.1	20.3
Cu-Ag / Ag-Cu	21.6	16.7	23.7	17.0	20.4	17.0
Cu-Ni / Ni-Cu	39.8	28.6	45.1	34.7	34.1	30.4
Cu-Ni-Sn / Sn-Ni-Cu	17.5	13.1	18.8	16.5	17.8	13.6
Al / Al	38.2	35.7	33.7	31.0	38.7	36.2
AlMgSi / AlMgSi	37.2	30.3	38.8	36.1	36.3	35.9

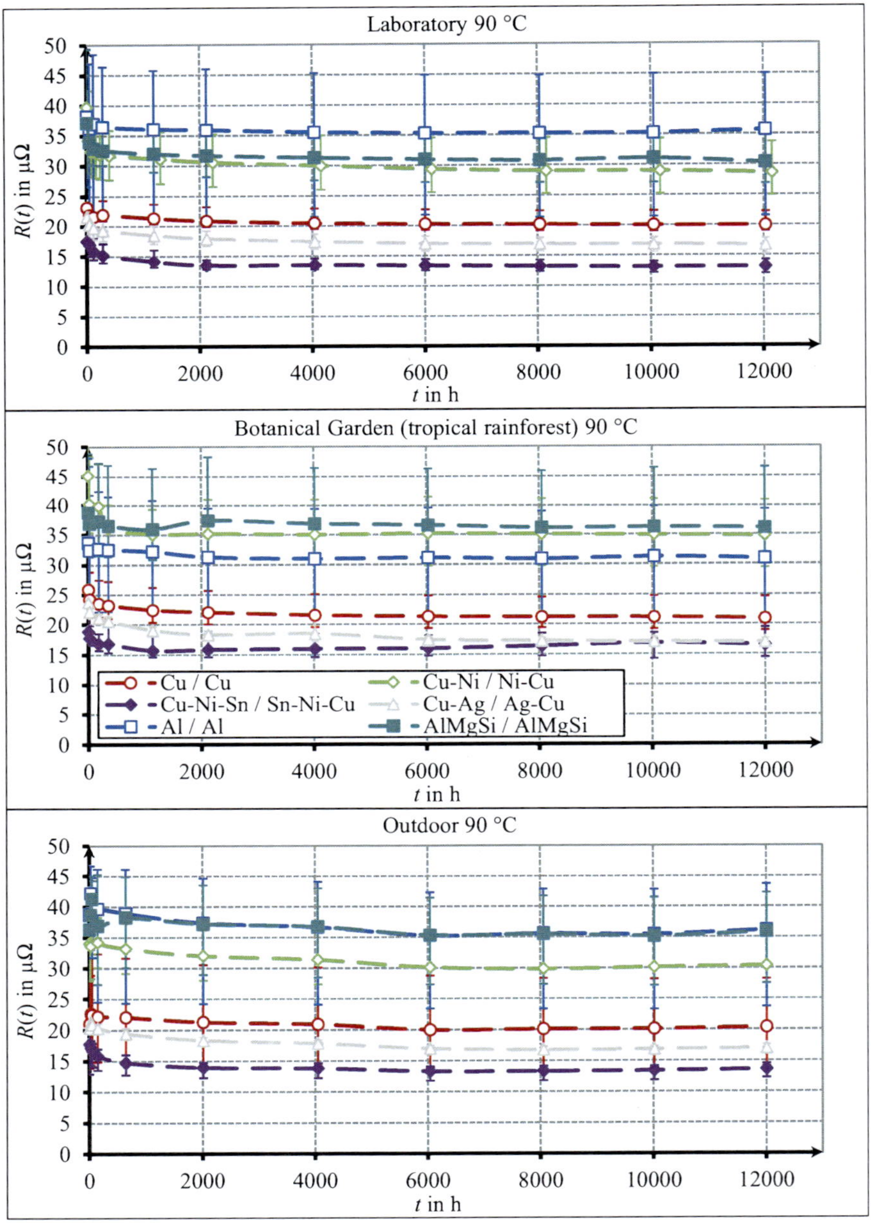

Fig. 42: Measured contact resistance dependent on time, temperature (90 °C) and different environments; resistance measurement was performed at ambient temperatures with a microohmmeter LoRe mobile (Werner Industrielle Elektronik, Appendix E)

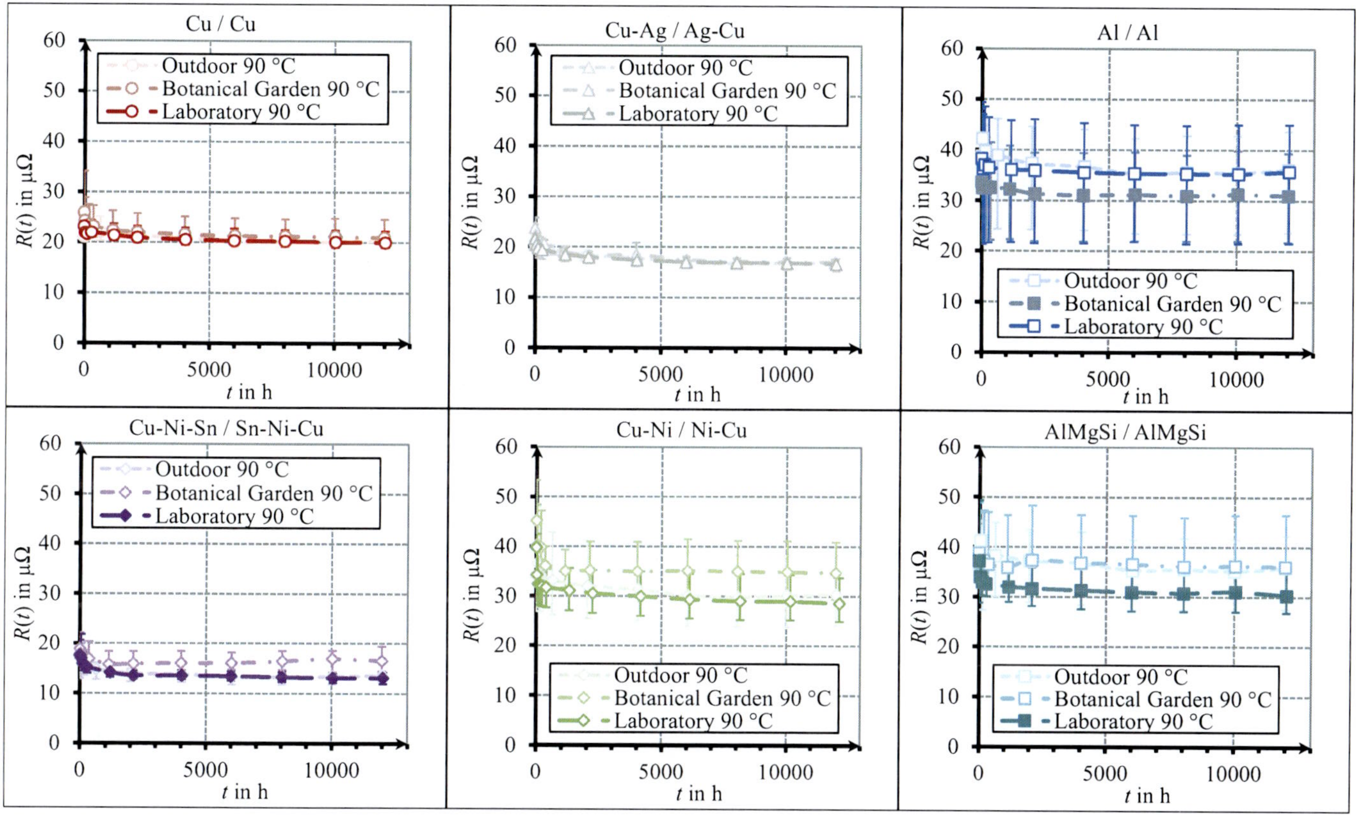

Fig. 43: Measured contact resistance dependent on time, material and coating and temperature (90 °C)

After 12000 h the contacts were opened and the radii of the plastic deformed area were determined and compared (Tab. 15). The outcome of the radii for each environment at 90 °C operating temperature is similar. Therefore, for $t = 12000$ h only the laboratory environment 90 °C is considered for comparison in the table below.

Tab. 15: Mean measured radii entire plastic deformed area $r_t(t)$ for $t = 12000$ h (≈ 1.4 a) at 90 °C and initial conditions $t = 0$

	Initial state (radii from Fig. 27)	Laboratory 90 °C
	$r_t(t = 0)$ in µm	$r_t(t = 1.4\text{ a})$ in µm
Cu / Cu	351	427
Cu-Ag / Ag-Cu	374	480
Cu-Ni / Ni-Cu	367	507
Cu-Ni-Sn / Sn-Ni-Cu	548	949
Al / Al	527	886
AlMgSi / AlMgSi	428	611

After 12000 h the radii (Tab. 15) significantly increase compared with initial state. Thermally activated material yielding occurs during operation. The springs ensured a constant force and avoided force reduction. The larger radius of Cu-Ni at 90 °C compared with Cu and Cu-Ag is partly reasoned by the radii determination method. The optical detection proceeding rests on a one-dimensional picture of the deformed area in axial direction. The deformation in axial direction is probably larger for Cu and Cu-Ag than Cu-Ni but cannot be determined by optical microscopy. The true deformed surface of Cu-Ag and Cu might be larger. Moreover, it can be seen, that the tin coating on Cu-Ni-Sn is almost entirely displaced (Fig. 46) from the contact area. Only some fragments remain on the nickel sublayer. In the non-contact area possible intermetallics develop with no impact on the contact resistance. On each material system film growth occur and was visible in case of Cu, Cu-Ni-Sn as well as Cu-Ag whereas the film on Cu-Ni, Al and AlMgSi is transparent (Appendix H, Fig. 96).

6.6.4 Long-Term Behavior of Crossed Rod Contacts at 200 °C

The used experimental design is based on the same principle as in Chapter 6.6.1 but was modified with regards to the operation temperature (Appendix H, Fig. 97). In Fig. 44 the plotted contact resistances are temperature corrected according to Eq. (30) and compared with the results of 90 °C in laboratory in the same graph. It can be seen from Fig. 44 that the measured resistance of all material systems significantly decreased dependent on time. The mean reduction of the measured resistance $R(t = 1.4\text{ a})$ based on $R(t = 0)$ is between 17 and 54 % dependent on material system. Nevertheless a thick film is visible on the surrounding surface of the contact area of Cu and Cu-Ag. The developed films on the surface of Cu-Ni, Al and AlMgSi were not distinctly visible by optical microscopy (Appendix H, Fig. 99 and Fig. 100).

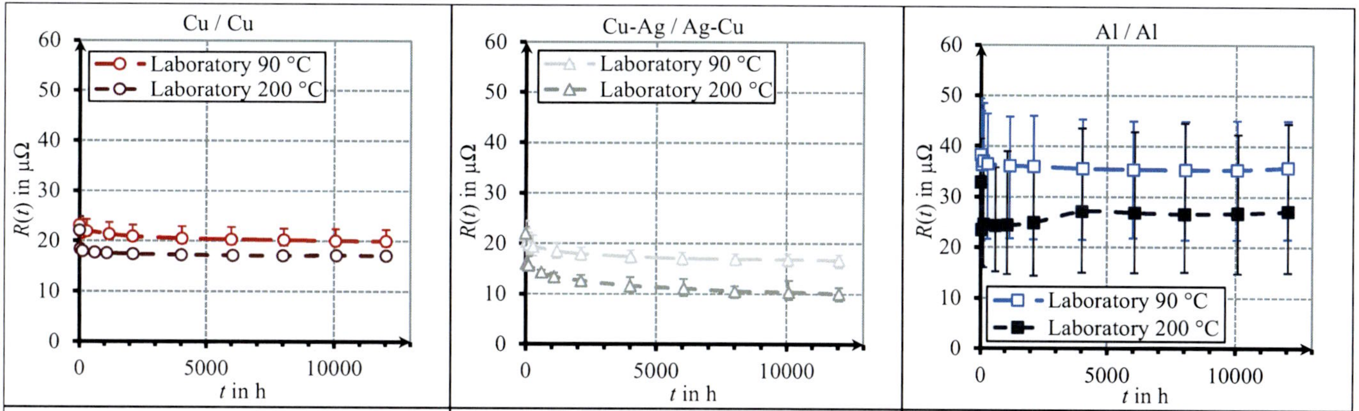

Mean measured resistance $R(t)$ for $t = 12000$ h (≈ 1.4 a) at 200 °C and initial state $t = 0$

Environment	Laboratory	
Thermal load in °C	200	
	$R(t)$ in µΩ	
	$t = 0$	$t = 1.4$ a
Cu / Cu	22.1	17.1
Cu-Ag / Ag-Cu	22.0	10.0
Cu-Ni / Ni-Cu	35.1	21.9
Al / Al	32.8	27.0
AlMgSi / AlMgSi	38.9	17.6

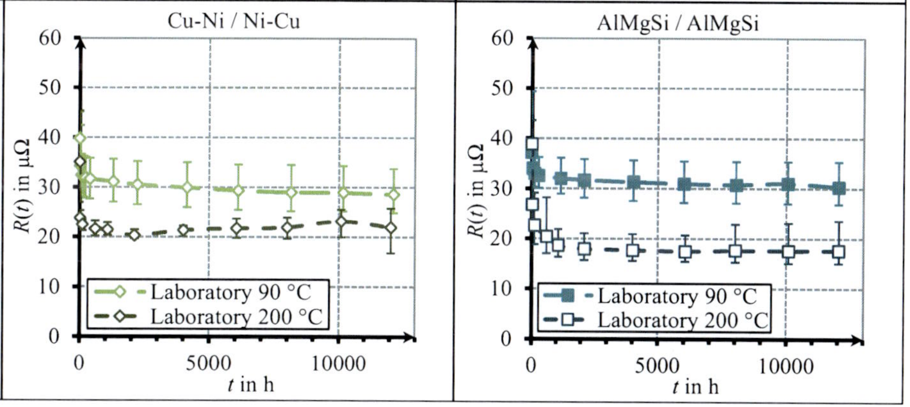

Fig. 44: Measured contact resistance dependent on time, material and coating for 90 °C and 200 °C

The experimental studies over 12000 h at 200 °C have shown that the principal long-term behavior is similar for 90 °C and 200 °C. In detail, some material and coating specific differences might exist. The resistance decrease in the phase of formation at 200 °C is significantly larger compared to 90 °C operating temperature. As a consequence, the contact area increases faster because of the temperature dependency of the yielding processes. Generally higher temperatures promote also the film growth. While the film growth is accelerated, the reversely yielding processes finally dominated the contact area increase in the formation phase. This process proceeds until the film on the periphery of the contact area cannot be penetrated by the contact partners (see also Chapter 6.5). The material yielding process itself continues but is decelerating. The constant force distributes over a larger area and results in decreasing mechanical stress. The remaining radii of the plastic deformed area r_t (t = 12000 h) are significantly larger after 200 °C operation than 90 °C (Tab. 15 and Tab. 16).

Tab. 16: Mean measured radii plastic deformed area r_t (t) at t = 12000 h (≈ 1.4 a) with 200 °C and initial conditions t = 0

	Initial state (radii from Fig. 27)	Laboratory 200 °C
	$r_t(t = 0)$ in µm	$r_t(t = 1.4$ a$)$ in µm
Cu / Cu	351	874
Cu-Ag / Ag-Cu	374	936
Cu-Ni / Ni-Cu	367	1010
Al / Al	527	2116
AlMgSi / AlMgSi	428	1351

Even Al presented the largest deformed plastic area after 12000 h, the relative reduction of contacts resistance is the lowest for AlMgSi. The softer Al is not able to penetrate the hard oxide film of aluminum oxide Al_2O_3. A slight increase of the measured resistance of Al was observed beyond reaching a minimum after the formation phase. It might be possible that oxide fragments are impressed towards the contact area. Subsequently after surface cleaning aluminum is covered with a thin oxide film. The current conduction occurs through small cracks in the oxide film. The constant contact force might possibly squeeze away very small a-spots from the cracks between present films in the contact area. This fact is difficult to prove and is superimposed with mechanical disturbance. Those disturbances are traceable to individual contacts and have a stronger impact on the contact resistance.

Cu presents the same long-term behavior as described in Chapter 6.5 for t = 1000 h. Similar to aluminum, the measured hardness does not significantly differ between t = 0 and t = 12000 h at 90 °C. However, the measured hardness for t = 12000 h and 200 °C is sufficiently lower (Appendix H, Tab. 27). In Fig. 45 the grain size changed dependent on time and temperature.

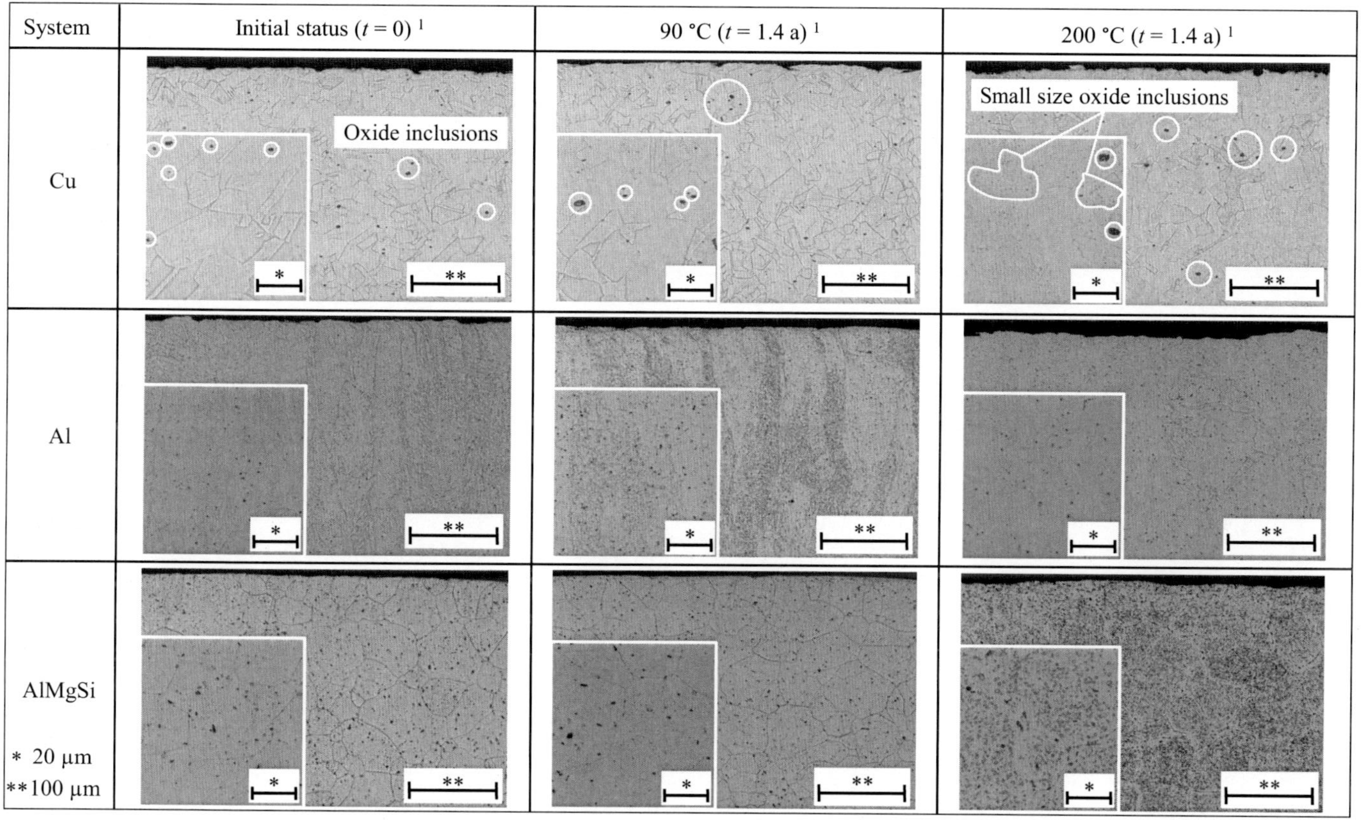

Fig. 45: Microstructure of Cu, Al and AlMgSi within the contact area dependent on time & temperature, [1] measurements done at IfWW, TUD

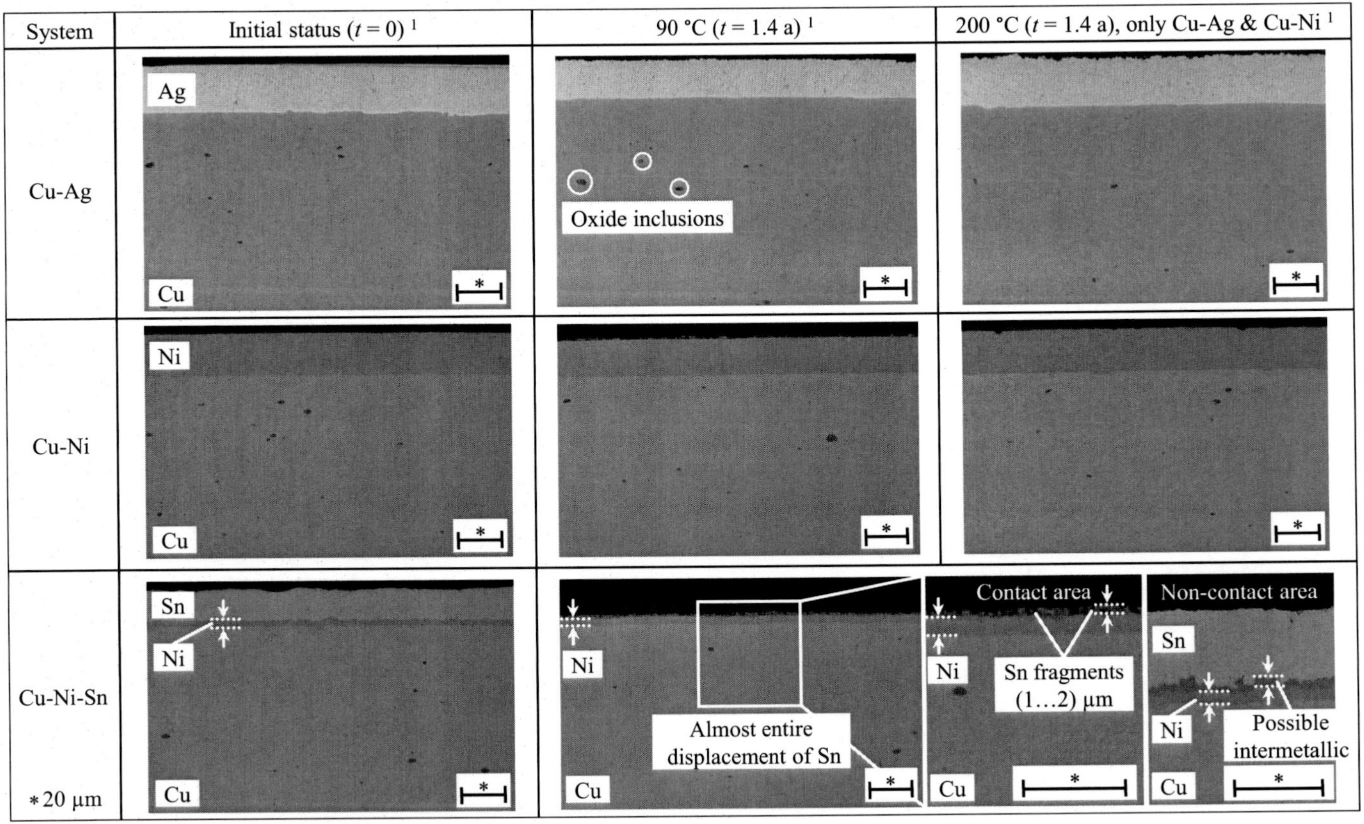

Fig. 46: Microstructure of Cu-Ag, Cu-Ni, Cu-Ni-Sn within the contact area dependent on time & temperature, [1] measurements done at IfWW, TUD

The grain boundaries in the microstructure of Cu, Al and AlMgSi (t = 12000 h, 200 °C) are less distinctive than at initial state or (t = 12000 h, 90 °C) because of possible recrystallization processes (see also Appendix H, Fig. 98). If not elsewise designate, all microstructure are virtually cross-sections of the contact area. The graph of Cu shows, that small size oxide inclusions have developed dependent on time and temperature. Compared with Cu and Cu-Ag, at Cu-Ni / Ni-Cu contacts larger radii of plastic deformed area (Tab. 16) were determined (see also Chapter 6.6.4 Cu-Ni at 90 °C) as well as the highest hardness of the considered material systems (Appendix H, Tab. 27). The measured hardness at t = 12000 h, 200 °C dropped about the half compared with the original state and t = 12000 h, 90 °C. The hardness of Cu-Ni is a mixture of substrate (Cu) and coating (Ni). Alternating coating thickness would have a remarkable impact on the hardness. Because of the different coating thickness (Chapter 6.1), more meaningful conclusions on the hardness of Cu-Ni are not possible. The increased scattering at t = 10000 h and t = 12000 h in resistance (Fig. 44) is reasoned by mechanical disturbance.

For the above treated material systems at 200 °C (Al, AlMgSi, Cu-Ni and partly Cu), the larger hardness of the native oxide and sufficient oxide film thickness are probably the cause that the contact resistance decline halts after the formation phase.

Fig. 47 shows that there has been a steady decline of the contact resistance of Cu-Ag / Ag-Cu over the entire operation time. Concordant with the other material systems the strongest resistance drop is within the formation phase. Even after the formation phase and t > 6000 the contact resistance declines.

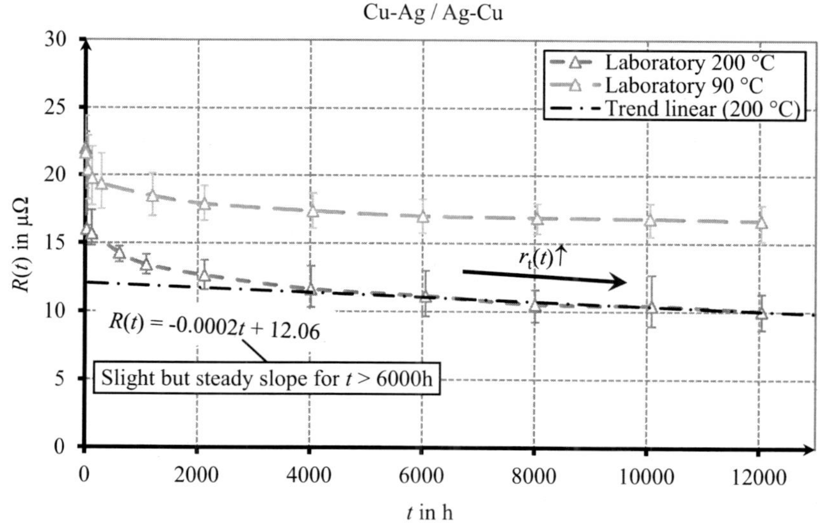

Fig. 47: Measured contact resistance Cu-Ag dependent on time

Possible films on the silver surface might be Ag_2O (when ozone is present) and Ag_2S (when hydrogen sulfide H_2S is present [53]). The permanent presence of Ag_2O is only theoretical because it is very soft, only weak adhesive and decomposes at 200 °C. The development of silver sulfide Ag_2S occurs already in the presence of very small amounts of H_2S in air. Ag_2S is not passivating and three times softer than silver itself [54]. Obviously Ag_2O or Ag_2S will be penetrated or squeezed away from the contact area by the constant force. This indicate that the contact area will increase as long as the force and the temperature are high enough. No equilibrium between film growth and material yielding will develop. The process is dominated by material yielding. However, for all considered environments film growth by chemical reactions is present. Nevertheless for force-stable stationary electrical contacts, the influence of films by chemical reactions may probably not result in significant contact degradation.

Recently, some research (unpublished work, TUD, 2015) indicated that diffusion of copper ions through the silver film could occur. The diffused copper ions may contribute to the development of copper oxide Cu_2O films on the silver surface. Typical colors of Ag_2S are lead-gray, gray or iron-black [55]. Thick Cu_2O films initially appear as red, then bronze, then brown and finally as black [27], [28], [32] (see also Fig. 37). Especially for thick films, the optical distinction between Ag_2S and Cu_2O might be challenging. In addition, both oxides may appear as mixed oxide on the surface of Cu-Ag systems. Furthermore the typical properties of such mixed oxide are not clear and needs to be studied or experimental determined.

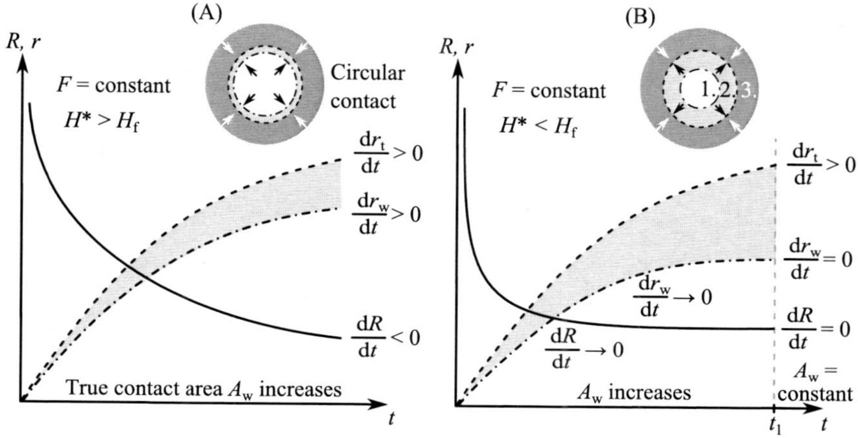

Fig. 48: General behavior of the contact resistance R dependent on time t and radial film growth for stationary electrical contacts applied in power engineering (with large force F = constant); H^* material hardness, H_f oxide film hardness, r_t radius deformed area, r_w radius true contact area or current-carrying area, A_w current-carrying contact area, t_1 time when dR / dt becomes zero, white arrows are the direction of radial film growth, black arrows are the direction of temperature dependent material yielding processes, 1. True contact area or current-carrying area, 2. Transition area, 3. Unimpaired film-covered area (insulating)

The finding of the studies in this chapter are concluded in the figure above. During operation temperature dependent material yielding processes and radial film growth occur. The yielding processes will increase the mechanical deformed area. If the contact material's hardness is higher than the oxide film ($H^* > H_f$) and F = constant, the film will be ruptured and metallic contact spots can develop (Fig. 48, (A)). As result the current-carrying contact area will increase. This process lasts as long as the force and the temperature are high enough. The slope dR / dt will reduce over time. The difference of the current-carrying contact area and the plastic deformed area depends on the oxide film properties. For an oxide with very weak mechanical properties and high electrical conductivity the radii $r_t \approx r_w$. Considering case (B), the slope of the radius of the mechanical deformed area is positive as long as force and temperature are high enough. The radius of the true contact area is increasing with declining slope until t_1. At time t_1 the increase in deformed area gives no increase in the true contact area because of the oxide hardness ($H^* < H_f$). The oxide film will be no longer ruptured. As result the slope of the resistance $dR / dt = 0$. The radial film growth will become very slow over time. While no mechanical disturbance or other impacts on the contact area occur, the contact resistance remains constant.

6.7 Long-Term Behavior of Stationary Electrical Connections at low Forces

The results of the long-term studies in Chapter 6.6 have shown that no significant increase of the contact resistance has occurred for stationary electrical contacts with a constant force. Considering the reference to the degradation mechanism chemical reactions, this finding is different to the results obtained by Takano [82] even the experimental design is not entirely comparable. Because of the experimental setup used in this work, some might resume that the results of Chapter 6.3 - 6.6 are reasoned by large contact areas and therefore very slow degradation progress. To prove this hypothesis, in the following chapter experimental results with cylindrical rods were conducted at suitable low forces.

6.7.1 Experimental Design and Setup

The experimental test were conducted in an oven with a temperature of 200 °C to promote film growth as described in Chapter 6.5. The studied materials are Cu and an AlMgSi. The material shape was extruded as cylindrical rods with diameter $d = 3$ mm. In Fig. 49, the designed experimental setup is shown.

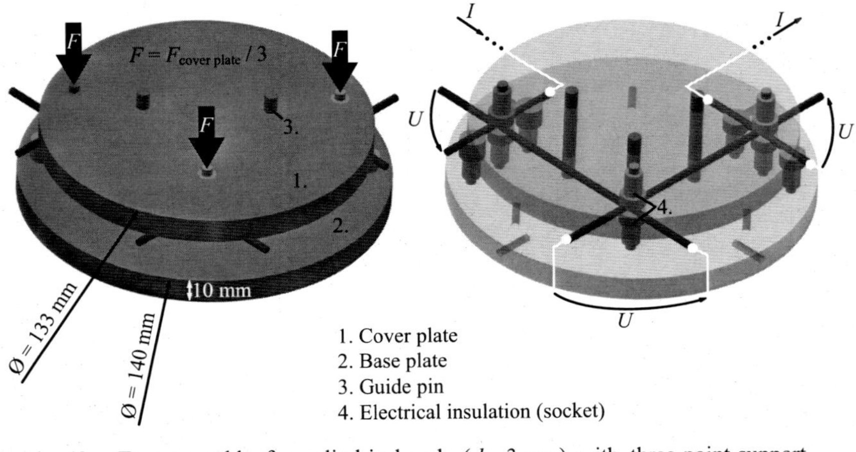

Fig. 49: Test assembly for cylindrical rods ($d = 3$ mm) with three-point-support system and low forces of ≈ 3.5 N per contact; the contact resistance was measured with a MI 3252 (Metrel, Appendix E)

The general test design is similar to that one described in Chapter 6.3 and 6.6 but is more specific to small rods and low forces. In Fig. 49, the test assembly contains three contacts of crossed rods. The rods rest in sockets made of polytetrafluoroethylene (PTFE). The rods are electrical insulated against the cover plate and the base plate by these sockets. Cover and base plate are made of stainless steel. The weight of the cover plate applies a mechanical load on the contacts. The position of the contacts is symmetrical around the center of mass of the cover plate. Therefore, each contact is exposed to nearly the same mechanical load. The load per contact is approximately ≈ 3.5 N and is one third of the complete mechanical load of the cover plate. During contact establishment the cover plate was carefully placed and lowered on guide pins. Nevertheless, small relative movements in non-vertical direction could not be completely excluded. A series of preliminary test was conducted to define load and dimensions as well as to ensure reproducibility. Before contact establishment, the rods were cleaned with ethanol, an abrasive pad (grit size 200 for Cu and 100 for AlMgSi) and a lint-free cloth. Four test assemblies with three contacts each were stored in an oven with a thermal load of 200 °C (Fig. 50). The rods and the wires for voltage taps and current input were mechanical decoupled by a ceramic luster terminal to avoid mechanical disturbance.

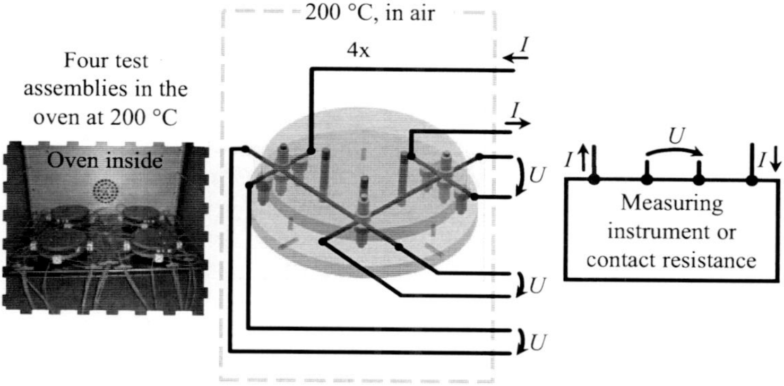

Fig. 50: Oven and experimental setup as well as test assemblies and devices, the contact resistance was measured with a MI 3252 (Metrel, Appendix E)

6.7.2 Procedure and Results

The resistance measurement was conducted regularly during operation with a low measuring current of 1 A to prevent additional heating. The measurements were temperature corrected to 20 °C with Eq. (30). During the operation time, the temperature was kept at 200 °C.

As can be seen in the figure below, a significant drop in resistance was observed in the formation phase. After passing the formation phase in the first 100 h no significant increase in resistance was observed. The range of the spreading and the resistance remain constant. The oxidation of the surface on the copper rods is as rapid as described in Chapter 6.5. The different areas and states of oxidation and deformation on copper are equally as mentioned in Chapter 6.5 and Appendix I. An equilibrium between thermally activated material yielding processes and film growth is established after the formation phase as illustrated in Fig. 38. A further test series with copper was conducted with a test duration up to 2000 h and the same result as described above (Appendix I, Fig. 102). The contact resistance for AlMgSi are constant in almost the same manner but whiteout visible oxide film. However, at the aluminum contacts cold welding occurred under the influence of the force and the temperature. This was observed when opening the contacts after 1000 h and can be partially seen on the microscopic images.

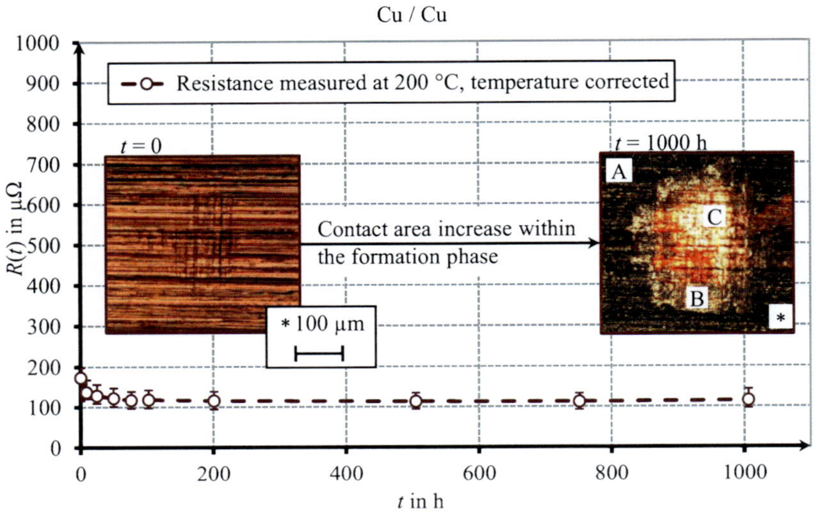

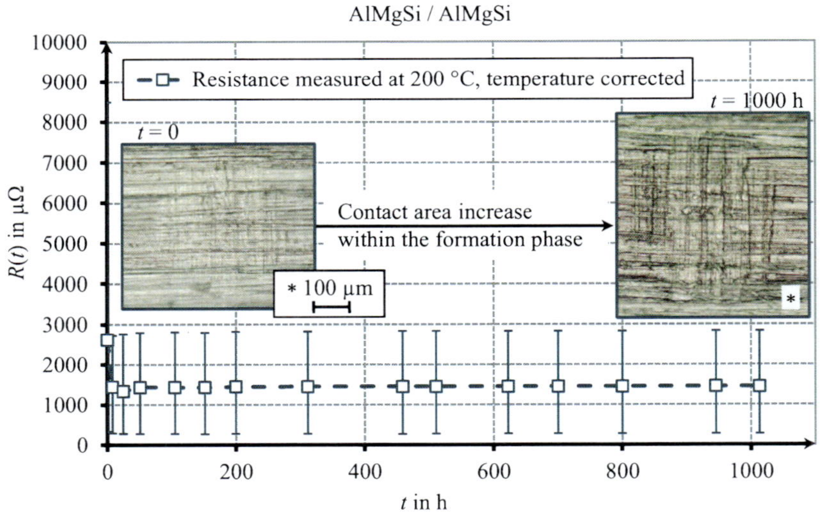

Fig. 51: Contact resistance of crossed rods (Cu and AlMgSi) dependent on time and temperature (200 °C) with exemplary microscopic images, temperature reference according to Eq. (30); Cu: A oxidized area with thick oxide film, B weak oxidized area, C deformation center

The outcome of the copper contacts is different to the findings of Takano [82]. However, he also reported decreasing contact resistance dependent on time for higher forces. Based on the findings of the study above, the observed degradation in other studies might possibly come from a) unwanted relative movements that may have a significant impact on the contact resistance. During the study above it was found that relative movements of the rods could shift the contact area in thick oxide film-covered insulating areas, regardless of whether this comes from thermal or mechanical causes. A further possibility could be, that b) a minimum contact force per contact spot exist, that prevent the spots from radial or vertical oxidation. It might be possible that as long as a constant force is present which is large enough to induce plastic deformation, no oxidation occur. For stationary electrical connections and contacts applied in power engineering, the forces are mostly high enough to result in plastic deformation of the contact surface. The continuity of the force application in such applications is determined by technical design and the material of the considered connection.

6.8 Impact of Anisotropic Film Growth on the Contact Resistance

Considering the film growth on the surface of an individual circular contact spot, only a few growth mechanisms seems to be possible. The film growth could occur in axial direction (*z*-direction, Fig. 52 (A)) on a plane contact area or in radial direction towards the contact area center (*x*-*y*-plane, Fig. 52 (B)). The combination of (A) and (B) is theoretically possible too.

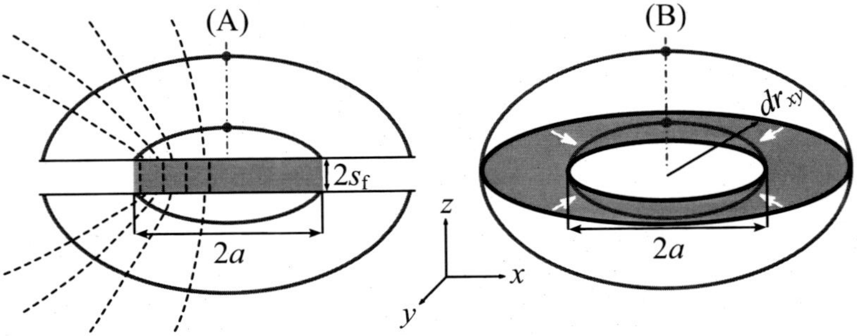

Fig. 52: Individual contact spot with axial (A) and radial (B) film growth, white arrows → direction of film growth

6.8.1 Contact Resistance and Axial Film Growth

Promptly after contact cleaning, a thin monolayer covers the metal surface. Charge carriers can pass these films by tunneling as long as the film thickness $s_f \leq (2\ldots5)$ nm. This film thickness is in the range of initially developed films after cleaning and depends on time, material and temperature. Natural developed oxide films may become thicker. Charger carriers could pass these films by common transport mechanism according to the band theory. It is assumed that axial developed films would significantly degrade the contact resistance.

Following, an individual *a*-spot of any bulk material ρ_b and spot radius a is considered where no films are present. The initial resistance $R(t = 0)$ is only determined by the constriction resistance: $R(t = 0) = \rho_b / (2 \cdot a)$. After aging with axial film growth, the same contact would be now determined by the constriction and an additional film resistance with film thickness s_f and resistivity ρ_f:

$$R(t) = \frac{\rho_b}{2 \cdot a} + 2 \cdot \frac{\rho_b \cdot s_f}{\pi \cdot a^2} \qquad (36)$$

The ratio $R(t) / R(t = 0)$ determines the relative degradation of the contact resistance. In terms of stationary contacts applied in electric power systems, a ratio $R(t) / R(t = 0) = 2$ would mean a significant degradation and $R(t) / R(t = 0) = 5$ probably the beginning of the thermal runaway. $R(t) / R(t = 0)$ can be transformed to obtain the film thickness s_f dependent on the axial film growth:

$$s_f = \left(\frac{R(t)}{R(t = 0)} - 1 \right) \cdot \frac{\rho_b \cdot \pi \cdot a}{4 \cdot \rho_f} \qquad (37)$$

The resulting film thickness s_f is determined by bulk ρ_b and film resistivity ρ_f, spot radius a as well as the determined ratio $R(t) / R(t=0) = (2\ldots5)$. According to the results in Chapter 6.3 and Appendix D considering multiple spots and circular cluster with radius a_c, a can be replaced by the radius of the plastic deformed area r_t. Finally, the described principle approach (see also Appendix J, Eqs. (52)-(54)) was used to solve the equations for all considered materials in order to determine the corresponding film thickness s_f (Appendix J, Tab. 29). The initial radii and the resistivity in Tab. 28 (Appendix J) were used to determine the theoretical thickness of axial developed films. Tab. 17 reveals that the obtained thicknesses are much lower than a monolayer of 0.1 nm (diameter of one atom) and seems to be physically impossible. Even for a possible thin layer of a few nm, the result of Eqs. (58), (62) and (66) would be a very strong discontinuity in resistance. This behavior is only theoretical possible and was not observed in this work or prior works with reference to the aging of stationary electrical contacts applied in electric power systems. It is obviously from Tab. 17, that significant axial film growth above (2…5) nm (boundary for tunneling) might not contribute to the continuous degradation of stationary power connections. The meaning of these results does not change for smaller radii or higher temperature (Appendix J, Tab. 28, Tab. 30). Only the resistivity of soft β-Ag_2S films at 200 °C might be low enough to enable significant current flow.

Tab. 17: Theoretical thickness of axial developed films dependent on the degradation $R(t) / R(t=0)$ according to Eqs. (55)-(66), temperature reference $\vartheta = 20$ °C

		Cu / Cu	Cu-Ag / Ag-Cu	Cu-Ni / Ni-Cu	Cu-Ni-Sn / Sn-Ni-Cu	Al / Al	AlMgSi / AlMgSi
		s_f in nm					
$\frac{R(t)}{R(t=0)}$	2	8.7E-10	1.3E-05	2.2E-15	7.6E-07	1.2E-18	1.2E-18
	5	3.5E-09	5.3E-05	9.0E-15	3.0E-06	4.8E-18	4.8E-18

6.8.2 Contact Resistance and Radial Film Growth

Radial film growth takes place at the boundary of an individual a-spot towards the spot center (Fig. 52 (B)). Possible film growth will reduce the metallic contact area. It is assumed that no significant current flow will occur in the film because of the weak conducting properties of possible oxide films.

In case of a large constriction, the transition at the contact interface from one conductor to another conductor can be assumed as very thin disc with thickness s_d. The disc is surrounded by equally thin oxide films (Fig. 52 (B)). The resistance of the thin disc R_d might be parallel to the resistance of the surrounding films R_f'. Both can be added to the resistance of the large constriction R_c': $R = R_c' + R_d \| R_f'$. The resistance of the disc is $R_d = \rho_b \cdot s_f / (\pi \cdot a^2)$. The individual resistance of the radial films would be either $\sigma_f / (\pi \cdot (r^2 - a^2))$ or $\rho_f \cdot s_f / (\pi \cdot (r^2 - a^2))$, dependent on whether if tunneling (tunnel film resistivity σ_f) or common transport mechanism ($\rho_f \cdot s_f$) are dominant. When tunneling is present, $R_d \| R_f'$ (Eq. (67), Appendix J) can be solved and transformed into Eq. (38):

$$R_d \| R_f' = \frac{1}{\pi \cdot \left(\frac{a^2}{\rho_b \cdot 2 s_d} + \frac{r^2 - a^2}{\sigma_f} \right)} \tag{38}$$

The equation is strongly dependent on $\rho_b \cdot s_d$ and σ_f. For $\rho_b \cdot s_d < a \vee \sigma_f < (r^2 - a^2)$, $R_d \| R_f'$ becomes small. The thickness of the disc is supposed to be very small. For the considered materials, the ratios $\sigma_f / \rho_b \cdot s_d$ and $\rho_f \cdot s_f / \rho_b \cdot s_d$ are large. Consequently, no significant current flow would occur through the radial developed films. The contact resistance would be determined by the constriction only. Certainly, radial developed films would reduce the current-carrying contact area. The constriction resistance R_f' might therefore depend on the radial film growth. For a long constriction with $\mu \to \infty$, Eq. (5) becomes to: $R_c'(t) = \rho_b / (2 \cdot a(t))$. The time dependency of the spot radius a can be expressed by linear, parabolic and other relationships dependent on material and temperature. This approach was already used in several models in more detail [37], [119], [120]. All of these models are very sensitive to parameter sets dependent on the model, including parameters such as temperature, frequency factor, activation energy, initial radius true contact area, mechanical stress and so on. Straight application of this models to determine $a(t)$ was not possible due to insufficient parameter assessment. All models state, that beyond the initial film development, the radial film growth might be dependent on the diffusion process of oxygen and metal atoms (Fig. 12). The diffusion process is very slow and ranges between several nm to a few μm within the considered models for aging time of several years [37]. It is questionable if these processes came into effect on the degradation of power connections with large contact area and many contact spots.

Theoretically, the radial film growth would proceed until $a(t) = 0$. However, the current would become more constricted and the current density J would increase. The higher constriction resistance would increase the resistive losses P according to $P = R \cdot I^2$ until the thermal runaway and failing of the contact. The finding on the degradation of multiple contact spots is similar because of the characteristic of the circular cluster (Appendix D).

6.8.3 Impact of Anisotropic Film Growth on the Current Distribution

Equally to Chapter 6.8.1 and 6.8.2, the impact of possible films can be distinguished in axial and radial developed films according to Fig. 52. Typical films at stationary electrical connections and contacts develop over several thousand hours. These films are relatively thick and mostly highly resistive. The current distribution in films with $\rho_f >> \rho_b$ is nearly uniform [1].

Radial films with $\rho_f >> \rho_b$ would result in a decrease of the true contact area and stronger current constriction. The outer rim of an individual *a*-spot is stressed with a higher current load than the contact area center. The local current load, dependent on the position, can be determined by the integration of Eq. (8):

$$I(r_{xy}) = \frac{I}{2\pi \cdot a} \int_0^{2\pi} \int_0^{r_{xy}} \frac{r}{\sqrt{a^2 - r^2}} dr d\varphi \qquad (39)$$

Solving Eq. (39) results in Eq. (40), (see also Appendix D, Eqs. (69)-(73)):

$$I(r_{xy}) = I - \frac{I}{a}\sqrt{a^2 - r_{xy}^2} \qquad (40)$$

The current I, is the entire current flow through the contact area $\pi \cdot a^2$ and r_{xy} is a variable radius within the contact area $\pi \cdot a^2$. Considering the influence of radial film development on the current distribution, the true contact area of an individual *a*-spot can be divided in several subareas.

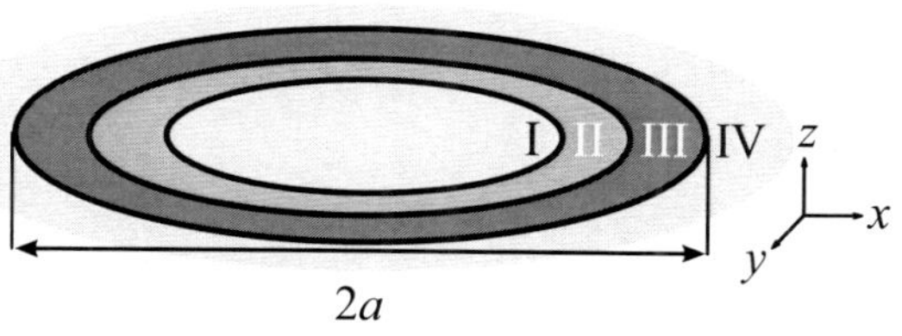

Fig. 53: Contact spot with zonal areas according to mean current density (I-III) and radial developed films (IV); I low, II medium, III high, IV zero

Tab. 18 reveals the mean current I_i and the mean current density J_i dependent on zone. The results show that the current density J_{III} is approximately $2.3 \cdot J_{II}$ and $2.8 \cdot J_I$. These findings indicate a significantly higher current load in area III, thus higher resistive losses. However, this might not came into effect because of the very effective thermal conduction in stationary power connections.

Tab. 18: Area A_i, current I_i and current density J_i dependent on radius r_{xyi} (see also Appendix J, Tab. 31)

	r_{xyi}	Area A_i	Mean Current I_i Eq. (40)	Mean current density $J_i = I_i / A_i$
I	$0 \ldots 0.5 \cdot a$	$0.25 \cdot \pi \cdot a^2$	$0.13 \cdot I$	$0.52 \cdot I / (\pi \cdot a^2)$
II	$0.5 \cdot a \ldots 0.75 \cdot a$	$0.31 \cdot \pi \cdot a^2$	$0.2 \cdot I$	$0.65 \cdot I / (\pi \cdot a^2)$
III	$0.75 \cdot a \ldots a$	$0.44 \cdot \pi \cdot a^2$	$0.66 \cdot I$	$1.5 \cdot I / (\pi \cdot a^2)$

According to $P = R \cdot I^2$, the resistive losses are $P \sim I^2$. The radial film development will result in a decrease of the current-carrying contact area and therefore higher contact resistance as well as higher current densities. In Tab. 32 (Appendix J) the contact area A_i, current I_i and current density J_i were determined dependent on zonal area and decreasing spot radius $a' = (0.01 \cdot a ... a)$. The determined current density J_i is plotted in Fig. 54. The results of Tab. 32 and Fig. 54 indicate that already a reduction of the spot radius of 30 % would result in a doubling of the current density. The ratios between contact areas A_i, current I_i and current density J_i will not change and remain constant for spot radius alteration.

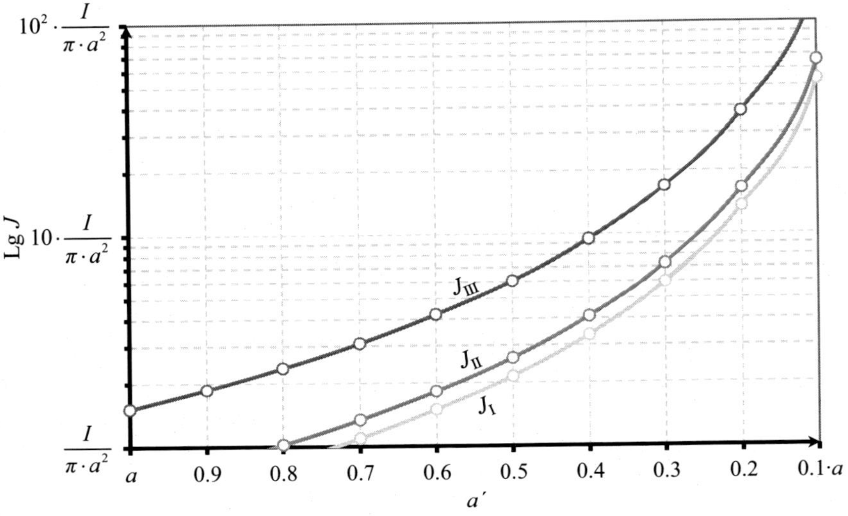

Fig. 54: Current density J depending on spot radius a'

The results suggest the theory that the equilibrium between Joule heating and heat conduction will not remain stable if contacts are going to fail. The area of stable thermal operating points would became smaller. Finally, the power balance of heat conduction and production would yield a surplus in heat production and therefore increasing contact temperature. The probability for contact failure might increase with increasing current density J.

Considering the case of multiple contact spots (Fig. 55), a circular cluster with radius a_c is assumed with a large number of n individual spots with radius a. The resistance of the entire contact might be determined by the circular cluster radius a_c (Appendix D).

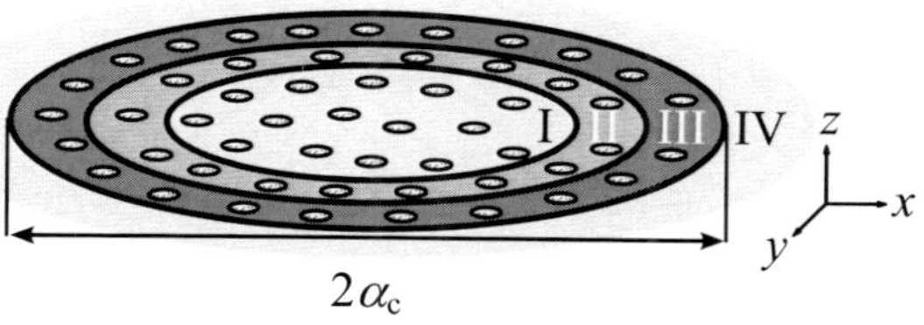

Fig. 55: Multiple contact spot with circular cluster radius α_c and zonal areas according to mean current density (I-III) & radial developed films (IV); I low, II medium, III high, IV zero

Similar to Fig. 53 the current distributes over the zonal areas. In [121]–[123], Malucci states that the principal current distribution on an exemplary circular multiple contact spot is similar to an individual spot. The highest current density occurs at the outer rim. He further mentioned a cascade effect of the degradation on the individual a-spots. If the spots on the outer rim would fail, the other spots will carry the additional current flow. This may become critical if the current density exceeds a specific level at which stable thermal operation is no longer possible. The determined ratios in Tab. 19 between area, current and current density remain nearly the same as in Tab. 18. In Eq. (40) a was replaced by α_c. In Tab. 34 (Appendix J) the degradation of the cluster radius α_c was determined. The ratios between the degraded areas, current and current densities remains also similar, thus the outcome is the same for individual a-spots and large circular cluster with radius α_c (Fig. 103). Consequently, it can be stated, that the findings on a circular cluster with radius α_c might be similar to that one for an individual a-spot. This is in accordance with [121]–[123]. However, these results were obtained for circular spots with large spot number n, typical for stationary power connections, and might possible change for contacts with lower spot number n and different shape. According to the findings in Chapter 6.6, 6.7 and 6.8.2 there is no distinct evidence whether the radial film development reduces the true contact area of stationary contact points with constant force. However, the outcome in this chapter on the current redistribution might be the same for other degradation mechanisms where the true contact area decreases dependent on time. Considering the results of Chapter 6.6, 6.7, other possible degradation mechanisms must be engaged to provoke significant contact degradation.

Tab. 19: Area A_i, current I_i and current density J_i dependent on radius r_{xyi}, (see also Appendix J, Tab. 33)

	r_{xyi}	Area A_i	Mean Current I_i Eq. (40)	Mean Current I_{ia} per spot	Mean current density $J_i = I_i / A_i$	Mean current density per spot $J_{ia} = I_{ia} / (\pi \cdot a^2)$
I	$0 \ldots 0.5 \cdot \alpha_c$	$0.25 \cdot \pi \cdot \alpha_c^2$	$0.13 \cdot I$	$0.52 \cdot I / n$	$0.52 \cdot I / (\pi \cdot \alpha_c^2)$	$0.52 \cdot I / (n \cdot \pi \cdot a^2)$
II	$0.5 \cdot \alpha_c \ldots 0.75 \cdot \alpha_c$	$0.31 \cdot \pi \cdot \alpha_c^2$	$0.2 \cdot I$	$0.65 \cdot I / n$	$0.65 \cdot I / (\pi \cdot \alpha_c^2)$	$0.65 \cdot I / (n \cdot \pi \cdot a^2)$
III	$0.75 \cdot \alpha_c \ldots \alpha_c$	$0.44 \cdot \pi \cdot \alpha_c^2$	$0.66 \cdot I$	$1.5 \cdot I / n$	$1.5 \cdot I / (\pi \cdot \alpha_c^2)$	$1.5 \cdot I / (n \cdot \pi \cdot a^2)$

6.9 Concluding Remarks

The film development caused by chemical reactions, is an important degradation mechanism for stationary electrical contacts applied in electric power systems. Corrosion processes could result in the loss of material and an additional force reduction. Common film development, such as oxidation, is often stated to be the reason for contact degradation during life time. In this research, the long-term behavior of stationary electrical contacts with crossed rods was investigated. The studied materials are Cu-ETP (bare or coated with, tin, silver or nickel), Al99.5 and AlMgSi0.5.

The applied contact force results in nearly circular contact points and mostly plastic deformation. The stress distribution within the contact area was determined with a FEM model. No significant difference of the stress distribution was observed in radial direction or in vicinity of the contact area. The FEM analysis shows a similar stress distribution for the region of the contact point and lies slightly below the ultimate tensile strength.

In experimental studies performed in ambient air, the contact resistance was measured at initial conditions. The measured contact resistance of Cu / Cu, Cu-Ag / Ag-Cu, Cu-Ni-Sn / Sn-Ni-Cu and Cu-Ni / Ni-Cu decreased with increasing force because of larger contact areas and also the rupture of possible films. The resistance of Al and AlMgSi contacts, measured in air, significantly spreads and was not in accordance with the analytical determined contact resistance $R_t(r_t)$. The significant difference between determined and measured contact resistance might be attributed towards the existence of non-ruptured high resistive aluminum oxide films within the contact area. Both, the studied material and the film influence, what kind of an impact the film might have on the contact resistance.

The film's influence on aluminum contacts at initial conditions was studied in an inert gas environment. Cleaning and the resistance measurement was done in an inert gas atmosphere (N_2) with a max. oxygen concentration of 0.03 % compared to ambient air (standard conditions). Distinctly lower resistances as well as a lower spreading were measured in N_2 for aluminum contacts. For pure aluminum, Al, the measured resistance in N_2 is approximately 40 % lower compared to the one measured in air.

In order to determine the impact of film development during the formation phase, tests with crossed copper rods were conducted in an oven at 200 °C for 1000 h. There is a clear trend of decreasing resistance within the first 100 hours. Afterwards, the resistance remains stable. The surface oxidized initially after the thermal operation commenced. After 1000 h the surface is covered by a thick apparently black oxide film. Microscopic images reveal that the entire plastic deformed area consists of several subareas with different grades of film development in radial direction towards the contact area center. The thermally activated material yielding process enlarged the plastic deformed area and therefore the true contact area. This process continues until the surface film becomes thick enough to no longer be penetrable. At this point, the increase of plastic deformed area continues but does not contribute to the true contact area. The contact resistance remains stable. These results suggest that an equilibrium was reached between material yielding and the oxide film growth process. The film growth already influences the contact resistance during the formation phase and restricts the decrease of the resistance by thermally activated material yielding processes.

In order to determine the film's influence on the long-term behavior, experimental tests with crossed rods were carried out in different environments for 12000 hours. The influences of other degradation mechanisms were suppressed due to the experimental test design. No significant resistance increase was observed for any of the considered environments at 90 °C operating temperature. In the formation phase, the measured resistance drops for all selected materials systems at 90 °C, even for aluminum (Al and AlMgSi). Eventually all contacts reached a stable resistance after 12000 h. The measurements on the contact resistance and the radii of the plastic deformed area indicate that the results may be similar for the different environments.

In experimental studies of crossed rods conducted at 200 °C for 12000 hours, the measured resistance of all considered material systems significantly decreased in the formation phase, even for aluminum, dependent on time. The resistance decrease during the formation phase at 200 °C is significantly larger compared to 90 °C, mainly because of the temperature dependency of the yielding process. The yielding process dominates the true contact area increase in the formation phase. This process continues until the film at the periphery of the contact area cannot be penetrated by the contact partners. However, the material yielding process itself and the increase in the plastic deformed area will continue. The processes decelerate due to the increasing contact area. After the formation phase has finished, the contact resistance's decline of Al / Al, AlMgSi / AlMgSi, Cu-Ni / Ni-Cu and partly of Cu / Cu halts when the native oxide film, characterized by a higher hardness and a sufficiently high thickness, cannot be penetrated by the yielding material anymore. For the contact resistance of Cu-Ag / Ag-Cu, a steady decline was observed over the entire operation time of 12000 hours at 200 °C. Films that might have evolved on the silver surface will be penetrated or squeezed away from the contact area. This indicates that for soft films, the true contact area will increase as long as forces and the temperatures are high enough. The yielding process will decelerate because of the larger contact area and therefore the lower mechanical stress, both dependent on time. The experimental study for 12000 h showed that the fundamental impact of the film development on the long-term behavior is similar at 90 °C and 200 °C operating temperature.

Film growth caused by chemical reactions occurs in all test environments. Nevertheless, it can be concluded that the films development without material losses on force-stable stationary electrical contacts may probably not result in significant contact degradation. This result was observed for contact points of crossed rods. It might differ for other power connections and will alter if corrosion processes take place in combination with a loss of material and a reduction of contact force.

The results above indicate that no significant increase of the contact resistances have occurred for stationary electrical contacts with a constant force. It could be presumed that large contact areas are the reason for a very slow aging progress. Supplementary studies with crossed rods ($d = 3$ mm) at very low force of 3.5 N per contact point were conducted in an oven at 200 °C for 1000 h. The observed contact resistance behavior for Cu / Cu and AlMgSi / AlMgSi contacts is almost in the same manner as the results presented above. The outcome of this study is different to other studies in the literature. This research work demonstrates that at crossed rods, even at low forces and small contact points, no contact resistance degradation occurs by film development. It was further observed that relative movements can have a significant impact on the contact resistance, which in turn could shift the contact member's true contact areas in thick oxide film-covered insulating areas. Below a threshold of 3.5 N, a possible minimum contact force per contact spot which prevents the spots from radial or vertical oxidation might exist.

Moreover, the theoretical anisotropic film growth was considered on circular contact points, either on individual or on multiple spots. Possibly, the film development above the tunnel film thickness could occur in axial or in radial direction towards the contact area center, or in both. When axial film growth occurs, the contact resistance can be determined by the constriction and an additional film resistance with film thickness s_f. The theoretical film thickness of non-tunneling films dependent on axial film growth was determined to be lower than a monolayer of 0.1 nm to result in significant resistance degradation. Therefore, axial film growth (non-tunnel films) might not contribute to the continuous aging of stationary power connections, even for smaller radii or higher temperature. A theoretical current flow might occur only at β-Ag_2S at 200 °C due to the low resistivity. However, β-Ag_2S is very soft and can be easily penetrated. Radial film growth takes place at the contact area boundary and will reduce the metallic contact area. Current flow through radial developed films can occur through very thin films by the tunnel effect. The current flow might probably not occur in parallel non-tunnel films as long as metallic contact spots exist. The constriction resistance might therefore be dominated by a time dependent contact area radius that can be described by time relationships dependent on material and temperature. Beyond the initial film development, radial film growth depends on the diffusion process of oxygen and metal atoms. As films will become thick and highly resistive, no current flow through the film might occur while $\rho_f >> \rho_b$.

7 Stationary Electrical Connections Operated in Harsh Environments

The application of stationary electrical connections and contacts in areas with harsh environmental conditions mean higher stress for the used connections and contacts. However, the performance and lifetime requirements still have to be fulfilled. The purpose of this study [124] is to investigate the long-term behavior of stationary electrical connections such as bolted busbar joints, in areas with prevailing harsh environmental conditions. Areas with harsh environmental environments, such as desert or tropical rainforest, can be characterized by high values in the ambient temperature ϑ_{amb}, ambient temperature spread ΔT_{amb}, relative humidity RH and reaction accelerators (gas components). The following field tests were conducted within such environments.

Tab. 20: Classified extreme conditions [77]

Designation	ϑ_{amb}	ΔT_{amb}	RH	Reaction accelerators (gas components)
Desert	↑↑	↑↑	↑	→
Tropical Rainforest	↑	→	↑↑	↑

It is supposed that electrical connections and contacts applied to such conditions would underlie enhanced degradation, especially film development caused by chemical reactions. Generally speaking, the aging process of stationary electrical connections and contacts applied in electric power systems can be divided into three phases:

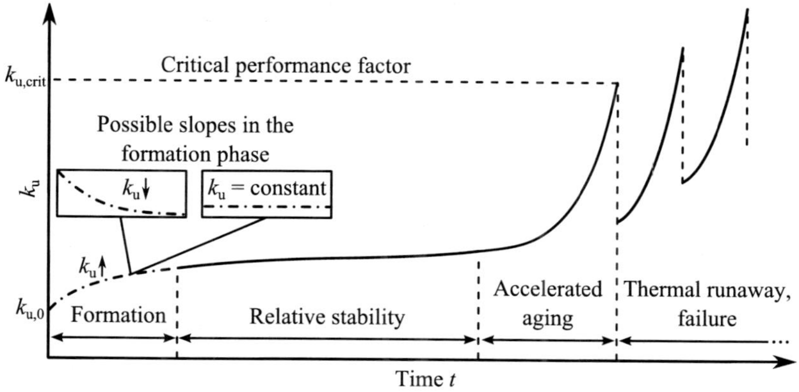

Fig. 56: Performance factor k_{u} dependent on time t for typical stationary electrical connections and contacts applied in electric power systems according to [76], [78], [125]

Within the formation phase (Fig. 56), initial physical processes, such as creep, thermal induced softening and embedding or stress relaxation take place. All of them influence the size of the current-carrying area A_{w} and therefore also the initial performance factor $k_{\mathrm{u,0}}$. The nature of this change depends on the material, force, shape, dimensions, temperature and time. The aging then proceeds with a low but increased slope of the performance factor k_{u} during the phase of relative stability. This results in a temperature increase at the contact's interface. The temperature at the interface increase considerably if the current-carrying area A_w falls below a critical value at which the generated heat of the interface cannot be transferred anymore to reach a steady state. Then the aging process proceeds into the phase of the accelerated aging until a critical performance factor $k_{u,crit}$ is reached. If the temperature exceeds the melting point of the material, the contact spots at the interface will melt. New contact spots will develop subsequently. The melting stops when the conducting area A_w is large enough for the Joule heating to fall below a limit at which the melting temperature is reached. The whole process does not establish the initial performance factor such as within the formation phase. This process rather alternates at higher levels until the thermal destruction of the contacts is completed.

7.1 Experimental Study Design

The test objects of this study are bolted busbar joints (cross section area 40 mm x 10 mm, length 150 mm) made of pure aluminum EN AW-1350A (Al99.5), aluminum alloy EN AW-6060 (AlMgSi) or pure copper CW004A (Cu). Each busbar has two drilled holes (d = 13 mm, Fig. 104, Appendix K) within the apparent contact area. The busbars were delivered by a metal manufacturer. Further manufacturing was done at TUD according to the dimensions above. The selected dimensions are based on various research studies [12], [13], [79], [83], [88], [116], [117], [126] conducted at TUD and should enable the comparison of the results.

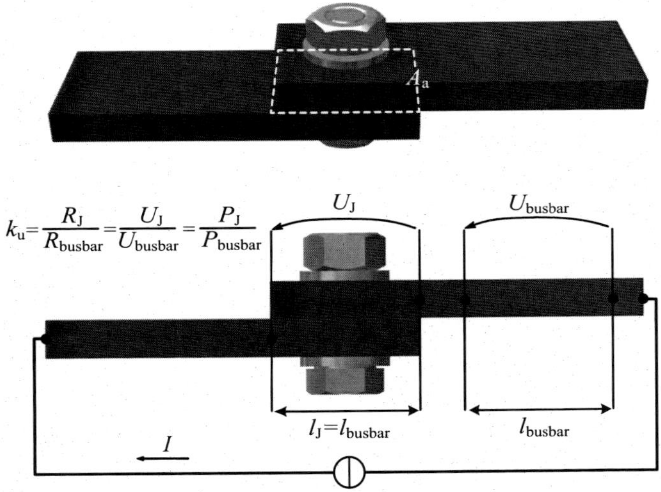

Fig. 57: Bolted busbar joint with apparent contact area A_a, performance factor k_u and schematic circuit; only one drilled hole per busbar is performed in the figure

The surfaces at the contact interface are either bare or plated with silver, tin or nickel. Before the connecting process, the busbar surfaces were cleaned by ethanol and with a wire brush (about 30 strokes with approximately 50 N compression force each) or an abrasive pad (grid size 200 and 1000) depending on the material and coating (Tab. 21) [12], [78]. In the field, inhibitors and lubricants are commonly used for switching-, plug-in-, sliding- and other moveable contacts. For some stationary connections such as press connectors, inhibitors are used for conservational issues and easier assembling. Using of inhibitors is not mandatory for force-fitted stationary connections, such as busbar joints. The long-term performance of bolted busbar joints might not be significantly improved by the use of inhibitors [88], [125]. Moreover, using inhibitors would require a detailed study on different inhibitors and its properties. The true influence chemical reactions might have on films would be blurred if inhibitors were utilized.

One material combination was set up with bare copper and aluminum alloy busbars using a single-sided aluminum-copper sheet (AlCu, $d = 2$ mm) between the two busbars. The aluminum-copper sheet consists of a cold rolled, single-side, composite material [127] ($d_{Al} \approx 1.4$ mm, $d_{Cu} \approx 0.6$ mm). After cleaning of the busbar surfaces, the busbars were bolted with galvanized or stainless steel M12x40-8.8 screws and M12 washers by using a torque of 70 Nm. Each busbar combination was assembled with five or more single samples. Most of the busbar joints were bolted on site while the joints in Brazil having partially been assembled in Dresden before transportation.

Tab. 21: Studied busbar material combinations

Base material / substrate	Coating (thickness + tolerance)	Designation[1]	Cleaning
Copper CW004A	Bare	Cu	Ethanol + wire brush
	Tin (6 + 3) µm Nickel (3 + 1) µm	Cu-Ni-Sn[2]	Ethanol
	Nickel (6 + 3) µm	Cu-Ni	Ethanol + abrasive pad, grit size 1000
Pure aluminum EN AW-1350A	Bare	Al99.5	Ethanol + wire brush
Aluminum alloy EN AW-6060 T6	Bare	AlMgSi	Ethanol + wire brush
	Silver (6 + 3) µm Nickel (3 + 1) µm	AlMgSi-Ni-Ag[2]	Ethanol
	Tin (6 + 3) µm Nickel (3 + 1) µm	AlMgSi-Ni-Sn[2]	Ethanol
	Nickel (6 + 3) µm	AlMgSi-Ni	Ethanol + abrasive pad, grit size 1000
CW004A, EN AW-6060 T6, aluminum-copper sheet (AlCu)	Bare, AlCu sheet	AlMgSi, AlCu, Cu	Ethanol + wire brush

[1] Connections are termed by material combination and slash " / " at the interface area, e.g. Cu / Cu (Chapter 7.2); substrate & coating are associated with a hyphen " - ", e.g. Cu-Ni

[2] Nickel is plated as thin sublayer to serve as diffusion barrier

The experimental research was carried out at the Botanical Garden of the Technische Universität Dresden (TUD) in Germany (DE), at the Suez Canal University Ismailia in Egypt (EG) and at the Universidade Federal do Pará in Belém, Brazil (BR). The busbars were bolted on site except the joints in BR. They were partially assembled before transportation. All samples were exposed to local environmental conditions. During the test duration environmental parameters, such as ambient temperature and relative humidity, were recorded (Fig. 58 and Fig. 59). Although the highest temperatures of > 45 °C as well as the highest spread between daily min. and max. temperatures of about 30 K was measured in Ismailia, EG. Fig. 58 shows that the lowest gap between min. and max. temperature for the tropical environment in Belém, BR was there. The relative humidity in this particular environment is very high and stays at a high level all year round. When the relative humidity exceeds 100 %, supersaturation of the air with fog formation and condensation occurs. Moreover, even in desert environments the relative humidity might be 100 %. During night time, the temperature decreases and the air cools down which results in a distinct dew formation in the morning hours of the following day, even in a dry environment.

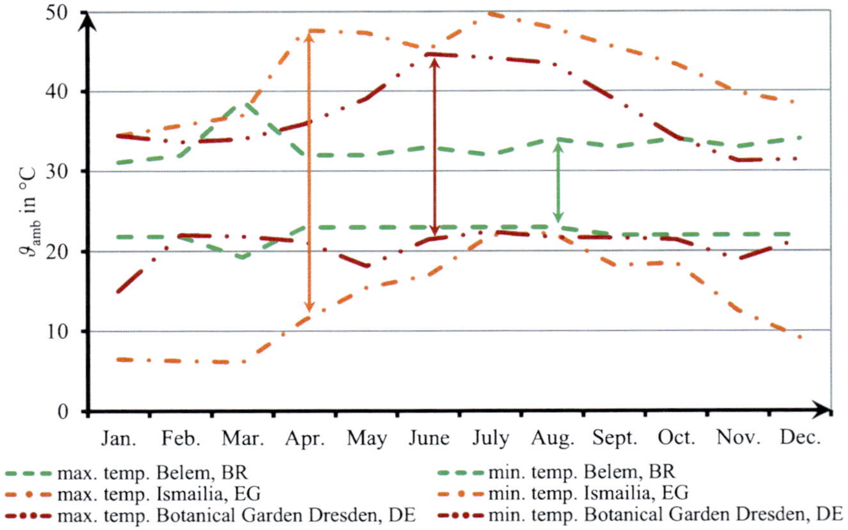

Fig. 58: Exemplary monthly min. and max. ambient temperatures ϑ_{amb} of different environments

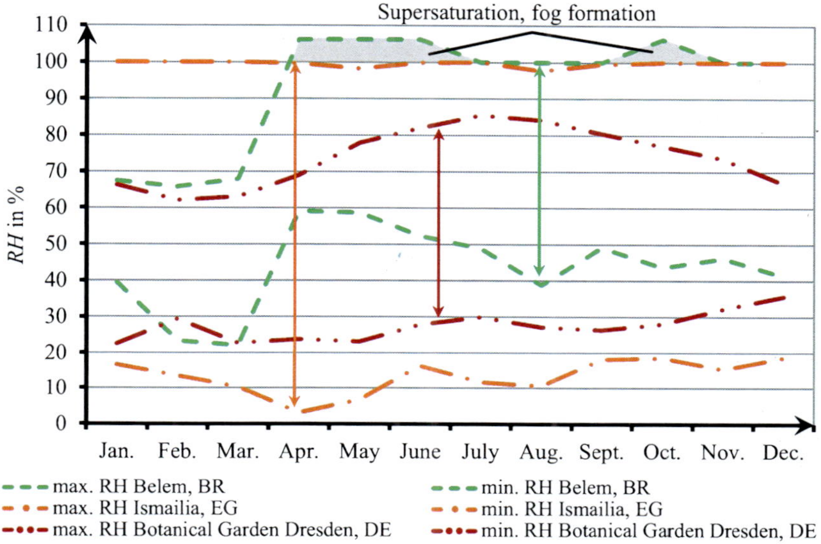

Fig. 59: Exemplary monthly min. and max. relative humidity *RH* of different environments

The experimental research was conducted both with current (EG, BR) and currentless (DE, BR). Test series with no current are indicated by supplement "currentless". As can be seen in Fig. 60 – Fig. 62, the busbar joints were connected in test circuits. A high current transformer RUE6500 So. 6.5kVA and a variable transformer RRSP2512 26-400-400-17 was used for each current-fed test circuit in BR and EG.

The permitted temperature of the considered bolted busbar joints depends on the material combination and the ambient temperature. According to IEC 62271-1:2007+A1:2011 and the used material combination, the permitted busbar joint temperature would be (90…105) °C for ambient temperatures of 40 °C. A max. operating temperature of 90 °C was set for comparability reasons between the connections. The current was adjusted as such to keep the connection temperatures at 90 °C. However, the daily change in weather variations affected the temperature of the tested connections. The temperature, measured at the connections in operation, was in the range of (30…90) °C in EG, (20…90) °C in BR and (15…45) °C in DE, dependent on local environmental conditions, such as day / night, rain, wind and further.

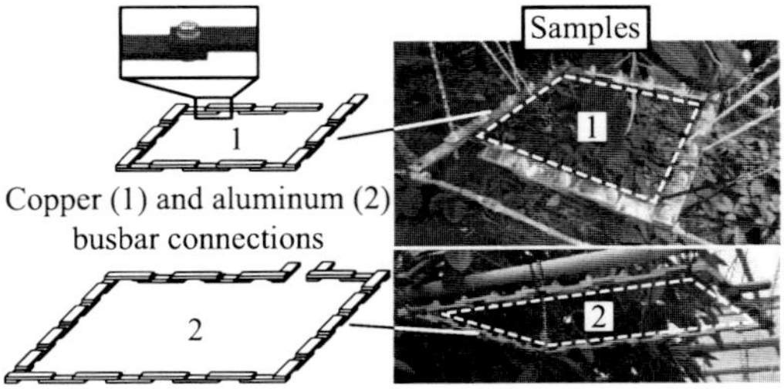

Fig. 60: Test setup Botanical garden TUD, DE

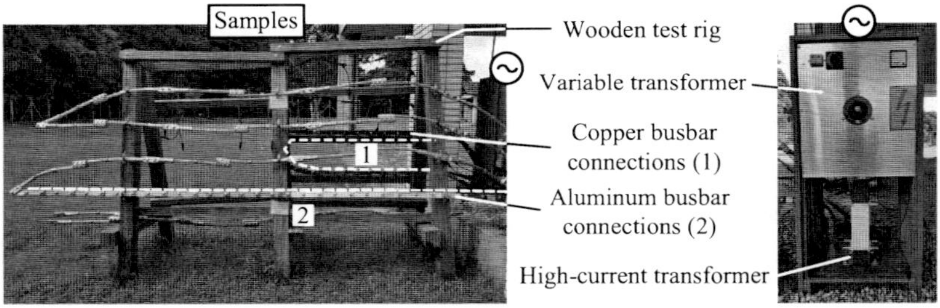

Fig. 61: Test setup Belém, BR

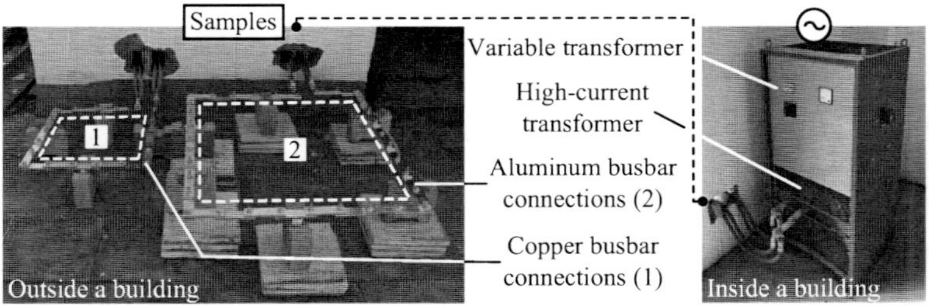

Fig. 62: Test setup Ismailia, EG

7.2 Long-Term Behavior of Busbar Connections

In order to obtain the resistance of a busbar sequence R_{busbar}, the resistivity of the used materials was determined. The resistivity (Tab. 22) was initially measured at ambient temperature and corrected to 20 °C with Eq. (30).

Tab. 22: Bulk resistivity of copper, aluminum and aluminum alloy used in power engineering at 20 °C, measurement done with a microohmmeter MO2A (Rasmus & Kühne, Appendix E)

$\rho_{20} = \frac{R_{20} \cdot A}{l} = \frac{R(\vartheta)}{1+\alpha_T \cdot (\vartheta - 20\ °C)} \cdot \frac{A}{l}$	Cu			Al			AlMgSi		
	$\bar{\mu}$	$s\uparrow$	$s\downarrow$	$\bar{\mu}$	$s\uparrow$	$s\downarrow$	$\bar{\mu}$	$s\uparrow$	$s\downarrow$
ρ_{20} in µΩ·mm	16.8	0.7	0.6	27.8	1.6	0.9	31.9	3.3	2.1

During the field work, the resistance measurement was conducted regularly with a microohmmeter LoRe mobile (Werner Industrielle Elektronik, Appendix E). Before the measurement was taken, the test circuits were switched off and opened at one position. A measurement current was fed into the circuit by the measuring instrument. Punch marks were used as potential taps (Fig. 57). For comparability reasons, the measured connection resistance was corrected to 20 °C with Eq. (30). The temperature of the busbar joints was measured together with the resistance at ambient temperature. The resistance results are in the range of a few µΩ. This is typical for the type of power connections tested.

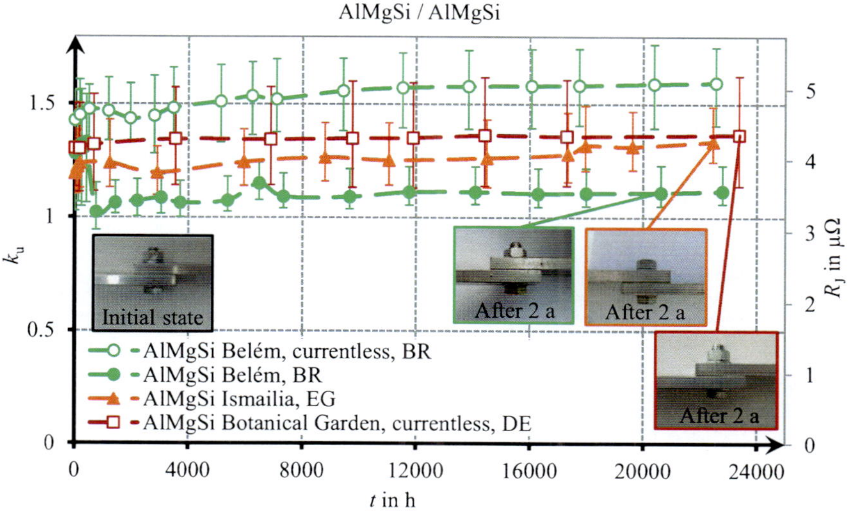

Fig. 63: Measured connection resistance of bolted AlMgSi busbar joints in different environments dependent on time, $\vartheta_{max} \leq 90$ °C

In the first 1000 h within the formation phase (Fig. 63), the measured resistance of some of the busbar joints rises. Both material and temperature depending processes provoke force reduction and smaller contact areas. The contrary drop for the current-fed samples in BR could be possibly caused by mechanical distortion. Higher initial connection resistance may result in enhanced aging [78]. The highest connection resistance for AlMgSi was measured for currentless busbar joints in BR (Fig. 63), possibly due to film formation. In addition to the native aluminum oxide layer found in humid atmospheres, a separate cover layer might develop and eventually further increases the film's thickness. Prevailingly high temperatures in the contact area would not only reduce the humidity locally but may also prevent a strong growth of such films. The connection resistances of the samples measured at the Botanical Garden are lower than the connection resistances of the currentless samples in BR. Possibly this might stem from the lower film's reduced impact by a lower mean relative humidity. The stronger scattering of the connection resistances measured at the Botanical Garden is caused by the simultaneous scattering of the ambient temperature. After being in the field for two years, the surface of the AlMgSi busbars (Fig. 63) appears somewhat matted. Meanwhile in the desert environment of Ismailia, EG, on the other hand, the surface is partially covered by dust particles.

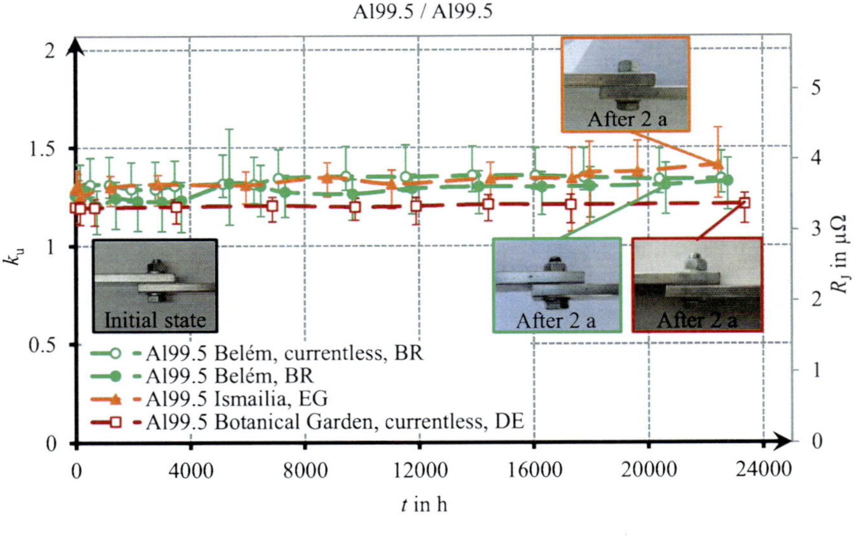

Fig. 64: Measured connections resistance of bolted Al99.5 busbar joints in different environments dependent on time, $\vartheta_{max} \leq 90$ °C

Compared with AlMgSi the difference of the measured connection resistances of Al99.5 (Fig. 64) in between the different environments are lower. There is no clear highest resistance graph based on the test environment. Eventually the spans overlap each other. Indeed, the mean connection resistances of Al99.5 / Al99.5 are lower than for AlMgSi. This a result of the better conductivity of Al99.5. However, the performance factor is nearly the same and indicates a similar quality of the electrical connection. No significant difference in performance factor of Al99.5 was observed between the samples with current or without current. It is therefore likely that the temperature influence on the considered operating parameters might be negligible for AlMgSi. After two years of experimental field work, the visual inspection of the busbar surface yields the same results as for AlMgSi (Fig. 64).

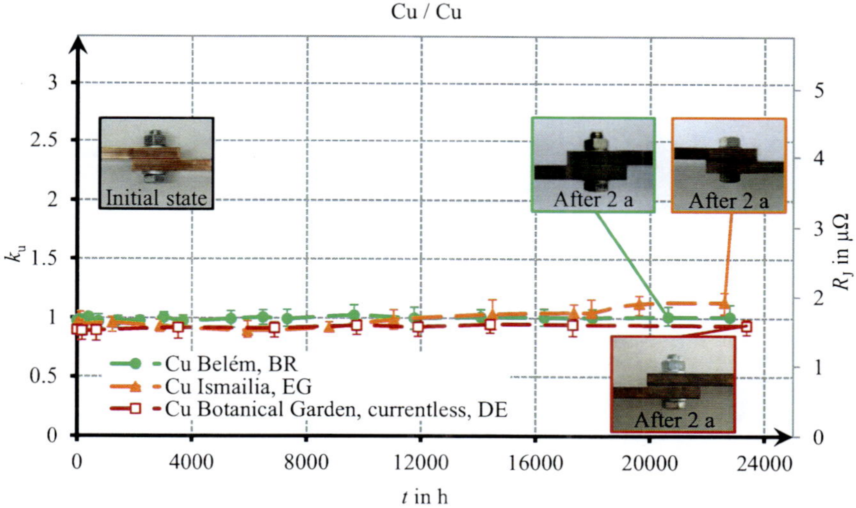

Fig. 65: Measured connection resistance of bolted Cu busbar joints in different environments dependent on time, $\vartheta_{max} \leq 90$ °C

The resistance of the bare copper joints (Fig. 65) remains stable over the duration of the test and it is also lower than that for aluminum. No visible difference was noticed among the different testing environments. A visible oxide film develops on the surface of the copper busbars (Fig. 65). The film appears darker for samples with current but also as tested in the tropical environment of Belém, BR. Temperature and humidity both influence the film development on the copper busbar surfaces. Nevertheless no influence on the connection resistance was observed.

It appears evident that the measured connection resistance of Cu-Ni-Sn / Sn-Ni-Cu and AlMgSi-Ni-Sn / Sn-Ni-AlMgSi joints (Fig. 66 and Fig. 67) only differ by the conductivity of the base material. Therefore, the measured connection resistances of AlMgSi-Ni-Sn / Sn-Ni-AlMgSi are slightly higher. No significant increase in resistance was observed during the long-term tests of tin-plated bolted busbar joints. The dimensionless performance factor is equal for both. After two years the surfaces have tarnished and changed color from bright to matt. Small frizzles and especially dust particles occurred on the surface.

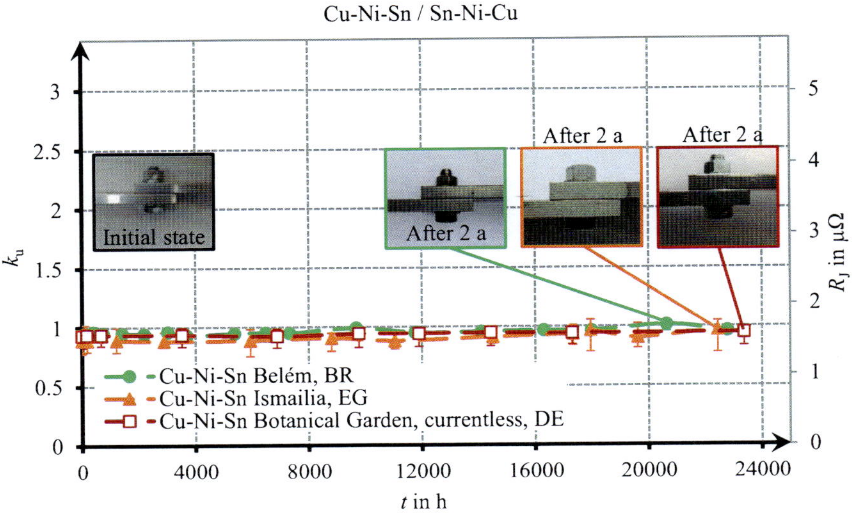

Fig. 66: Measured connection resistance of bolted Cu-Ni-Sn busbar joints in different environments dependent on time, $\vartheta_{max} \leq 90$ °C

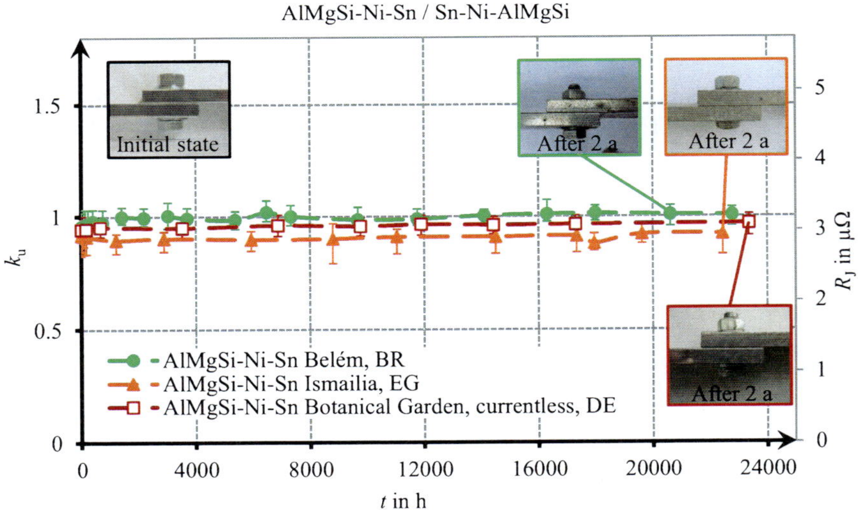

Fig. 67: Measured connection resistance of bolted AlMgSi-Ni-Sn busbar joints in different environments dependent on time, $\vartheta_{max} \leq 90$ °C

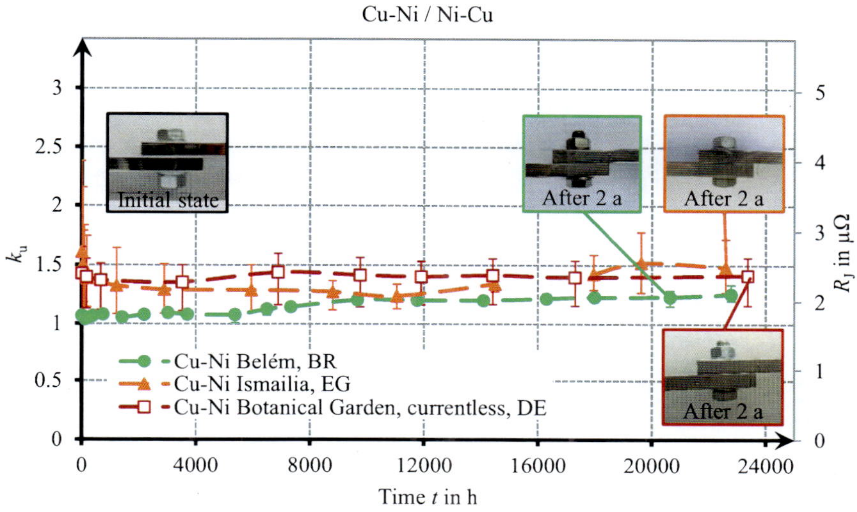

Fig. 68: Measured connections resistance of bolted Cu-Ni busbar joints in different environments dependent on time, $\vartheta_{max} \leq 90$ °C

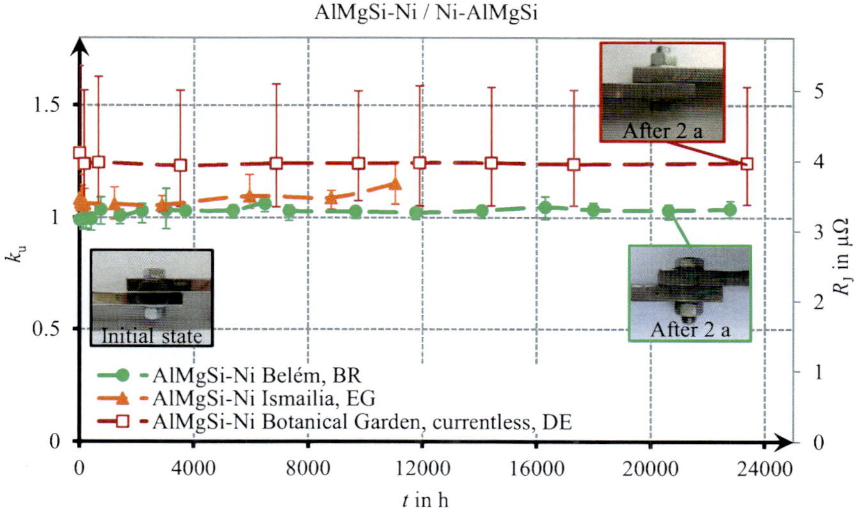

Fig. 69: Measured connection resistance of bolted AlMgSi-Ni busbar joints in different environments dependent on time, $\vartheta_{max} \leq 90$ °C

The measured connection resistance of Ni plated Cu and AlMgSi busbars (Fig. 68 and Fig. 69) are relatively close to each other. The higher connections resistance of AlMgSi-Ni / Ni-AlMgSi Botanical Garden, currentless, DE (Fig. 69) is caused by two individual busbar joints. Applicable to the samples used in Ismailia, EG, a decrease of the measured connections resistance can be observed during the first 1000 h of the formation phase. At this time, material and temperature-dependent processes produce larger contact areas and therefore lower connection resistances. The connections resistance of the AlMgSi-Ni / Ni-AlMgSi in Ismailia, EG (Fig. 69) was evaluated at approximately 11000 h only. Afterwards unwanted mechanical distortion occurred at the test site. Before assembling, the Ni coated busbars in Belém, BR were additional treated with an abrasive pad (grid size 1000). The reduction of the connection resistance during formation phase is therefore not perceptible for the Ni plated aluminum busbars in Belém, BR (Fig. 69). On the surface of the Ni plated busbars no clear visible film developed and the surface remained bright. Typical oxide films on nickel are very thin and may appear transparent. Appropriate surface cleaning is essential and could have strong influence on the connection resistance. This fact is evident when comparing the performance factor of Cu-Ni / Ni-Cu in Ismailia, EG to that of Belém, BR (Fig. 68 and Fig. 69). In Ismailia, EG and Dresden, DE, no additional surface cleaning using an abrasive pad was done prior to the tests. There the performance factor scatters stronger, which is induced by more or less penetrated films.

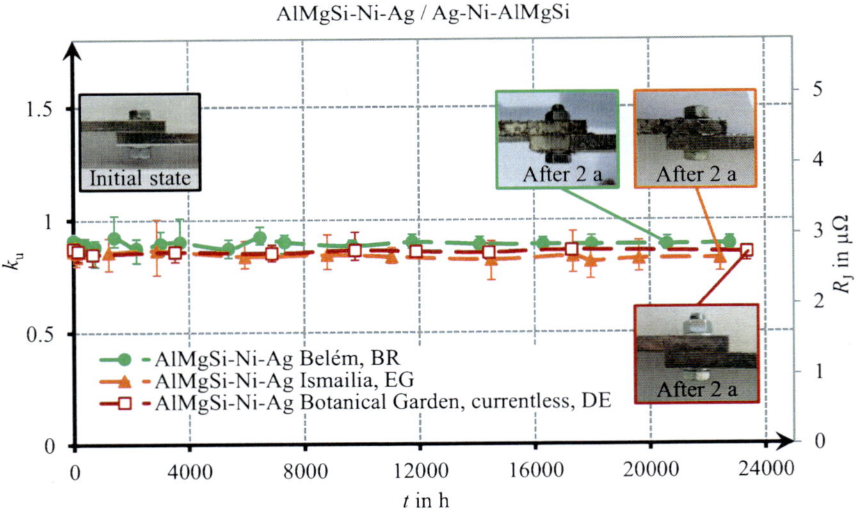

Fig. 70: Measured connections resistance of bolted AlMgSi-Ni-Ag busbar joints in different environments dependent on time, $\vartheta_{max} \leq 90$ °C

The connections resistance of AlMgSi-Ni-Ag / Ag-Ni-AlMgSi (Fig. 70) did not show significant changes during the whole testing duration. Considering the resistances only, no significant difference can be observed between the three environments. Although, the silver surface did show strong chances. Overall the busbar surface appears in colors ranging from bright to more dark and black with no uniformity in place. Moreover, the coating was locally spalling and heavily destroyed (Fig. 71 below). Possibly, leaky coatings enable layer infiltration. Despite this fact the performance factor remains stable; the most likely reason for this being that the coating at the interface area is compressed by the connection force between the busbars. As a result, the silver coating in the apparent contact area is somehow protected against environmental effects and the measured connection resistance also remains stable. Samples with current present a stronger environmental attack on the busbar surface because of the additional influence of the current-induced heating (Fig. 71, Ismailia EG, and Belém, BR). The higher the temperatures, the stronger the possible environmental attack.

Initial state	Dresden, after 2 a	Ismailia, after 2 a	Belém, after 2 a

Fig. 71: Environmental attack on the surface of AlMgSi-Ni-Ag busbars

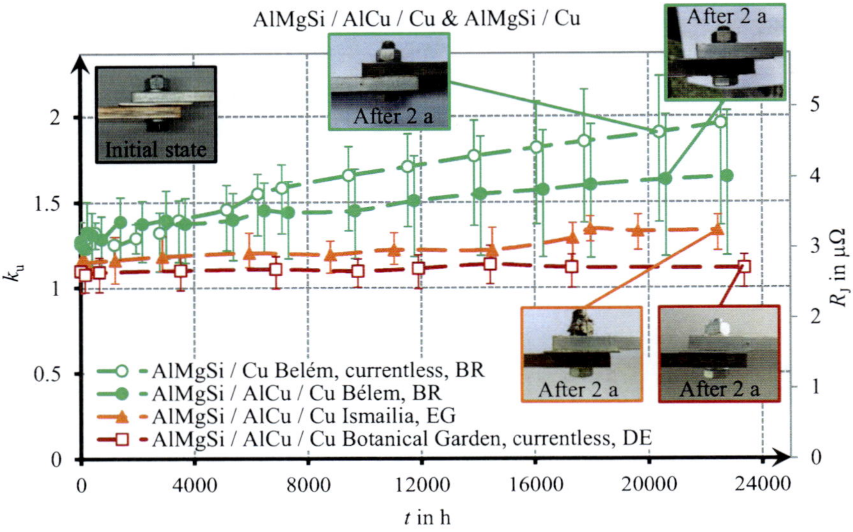

Fig. 72: Measured connection resistance of bolted AlMgSi / AlCu / Cu and AlMgSi / Cu busbar joints in different environments dependent on time, $\vartheta_{max} \leq 90$ °C

For the material combinations displayed in the figure above (except for the combination AlMgSi-Cu Belém, currentless, BR), AlCu sheets (AlCu) were used between the interfaces of the busbar surfaces (Fig. 72). The combination AlMgSi / Cu was tested in order to demonstrate the difference when using no AlCu sheet. The highest connections resistance graph was measured for the samples in Belém, BR. In Ismailia, EG and the Botanical Garden the connections resistances remained stable. Previous research stated that AlMgSi / Cu tend to enable electrochemical reactions between the materials [3], [7]. Possibly, the graph of AlMgSi / AlCu / Cu in Belém was not stable due to the same reason as applicable to AlMgSi / Cu. Maybe the electrochemical reaction occurring between aluminum and copper can also proceed along the surface of the AlCu sheet after all. A sufficiently humid atmosphere, as present in Belém, BR, would be required tough. In the future, the bolted AlMgSi / AlCu / Cu busbar joint in Belém will exceed the performance factor $k_u \geq 1.5$ and would convert from a heat sink to a heat source.

Depending on the environment, the visible surface alterations are the same as those for aluminum and copper. Additionally, a greenish and whitely film can be also seen at the interface area of the AlMgSi / Cu busbar joint in Belém, BR. More studies seems to be necessary to understand the failure mode of Al / Cu connections [86].

7.3 Concluding Remarks

The purpose of this research was to assess the long-term performance of stationary electrical connections dependent on time, temperature, material and the environment. It was designed to determine what effect, if any, the degradation mechanism chemical reactions might have on stationary electrical connections applied in real operating conditions.

This is one of the first experimental works investigating what effects, if any, the degradation mechanism chemical reactions has on the long-term behavior of stationary electrical connections (for power engineering) in tropical and desert environments all dependent on time (> 20000 h), material and temperature (current-fed / currentless).

This work has found that in areas with desert and tropical rainforest conditions, bare busbar joints of the same material, either made of copper, aluminum or aluminum alloy, show stable performance factors throughout the testing time duration of more than two years.

In tropical rainforest and desert areas, tin coated busbar joints based on copper or aluminum alloy (AlMgSi) present stable connection resistances and performance ratios in long-term tests conducted over two years.

The investigation of silver coated busbar surfaces has shown that the environmental attack is strong and leads to spalling and destroying of the silver coating. The significant presence of sulfur compounds in the surrounding air might be the reason for the strong surface alteration. However, the performance factor of silver coated aluminum busbar joints, exposed to tropical rainforest conditions, remains stable in long-term tests lasting over two years. One of the more significant findings to emerge from this study is that the silver coating in the contact area is somehow compressed by the connection force of the bolted busbar joints. The silver coating in the contact area remained electrical unimpaired and resulted in a stable connection resistance.

The research has also shown that the performance factor of nickel coated busbar joints exposed to desert conditions remains stable in long term tests up to one year. Longer test durations seem necessary for meaningful results. In tropical rainforest conditions, bolted busbar joints show a good performance in tests up to two years' time. Bolted busbar joints with nickel coatings are supposed to behave more sensitive to mechanical distortion than any other tested material combinations.

During the test duration of more than two years, the performance factor of bolted copper and aluminum busbars connected via AlCu sheet was stable in a desert environment and was rising in a tropical rainforest environment. The chemical reaction between aluminum and copper proceeds in very humid environments. Regarding tropical rainforest environments, the high amount of water in the air could possibly enable the chemical reaction to proceed along the surface of the AlCu sheet. More studies seem to be necessary to understand the failure mode of Al / Cu connections [86].

Finally, during the experimental phase of the tests, a low performance factor and no significant change of the connection resistance was observed. No distinct difference between the long-term behavior in harsh natural environments and laboratory tests was observed. According to prior works, it is expected, that electrical connections with $k_u < 1.5$ will not significantly degrade in the future [12], [76], [79], [83], [86]. The results of this study indicate that within the tested environments, the film development induced by chemical reactions may not be critical for this kind of power connections.

8 Conclusion

Chemical reactions on the surface of electrical connections and contacts result in the development of films. In the literature it is reported that such films have a significant impact on the long-term behavior of stationary electrical connections and contacts that is not fully understood as of yet. In this research work, the impact of film development caused by chemical reactions, such as oxidation, with no significant loss of material was studied dependent on time, material and the environment. Crossed rods with circular contact points were used in experimental, numerical and analytical studies.

Numerical analysis reveals that for the considered force when plastic deformation is dominant, the mechanical stress distribution within the contact area is approximately equal (Chapter 6.2). This is in accordance with prior works.

For electrical measurements in air (standard atmosphere), the contact resistance was determined together with the plastic deformed area. Good accordance between the results of the measurement and the analytical approaches was achieved for bare, silver, tin- and partially nickel coated copper contacts. The measured contact resistance of aluminum, either alloy or pure, is above the theoretical expectations and significantly spreads. This might be caused by oxide films. The film's influence on the contact resistance can be seen by measurements in low oxygen inert gas atmosphere which produced more homogenous results with lower contact resistance for aluminum contacts (Fig. 33, Chapter 6.3 and 6.4).

The film's impact on the contact resistance during the formation phase was investigated for copper contacts with constant force in an oven at 200 °C for 1000 h. While the operation of the test, temperature dependent material yielding processes and radial film growth occurred simultaneously. For the contact material hardness H^* that is higher than the oxide film's hardness H_f ($H^* > H_f$) with $F = \text{constant}$, the film will be ruptured and metallic contact spots might develop. This process will continue as long as the mechanical stress and the temperature are sufficiently high enough. After the formation phase, the increase of the deformed area does not result in an increase in the true contact area as the oxide will no longer be ruptured for contacts with oxide films ($H^* < H_f$). As a consequence, the slope of the resistance $\mathrm{d}R / \mathrm{d}t = 0$ (Fig. 48, Chapter 6.5).

The impact of film development dependent on time, environment, material and temperature (90 °C and 200 °C) was studied on crossed rods with constant force in long-term tests. No significant degradation was observed for the test environments laboratory (indoor), botanical garden (tropical rainforest) and outdoor at an operating temperature of 90 °C. The outcome of long-term tests on crossed rods at 200 °C for the laboratory setting was the same. However, in the long-term, the individual film development caused by chemical reactions (such as oxidation) might not result in a significant degradation of stationary electrical contacts with circular contact points and a constant force (Chapter 6.6). This fact was confirmed for experimental tests of crossed rods at 200 °C and very low forces (3.5 N per contact point, Chapter 6.7).

Through theoretical analysis, it was shown, that axial film growth above the tunnel film thickness would result in excessive resistance increase that was not observed in this study or previous work. Although radial film growth might degrade the contact resistance, its growth is probably too slow to provoke failure cause (Chapter 6.8).

No literature considering the long-term behavior of stationary electrical connections operated in harsh environments, such as desert and tropical rainforest, was found. Experimental field work was carried out to supplement prior laboratory works with the purpose of studying the long-term behavior of busbar joints dependent on time and material operated in harsh environments. Except nickel coated aluminum busbars, the most of the considered material combinations show stable performance factors. The strong environmental impact was seen on the surface of silver coated busbars (Fig. 71, Chapter 7).

Even if the plastic deformed area of plane contacts, such as bolted busbars, is larger, the above achieved results for circular contact points are not straightly applicable on plane contacts. This is reasoned by the nonuniform stress distribution for this type of power connections. However, the strong impact of oxide films on aluminum was observed. In addition to that, the limiting influence of the film growth on the contact resistance decrease during the formation phase was clarified. This research work has eventually shown that the film development induced by chemical reactions (such as oxidation) does not result in a significant time dependent resistance degradation of circular contact points with constant force within the considered testing time. In fact, this indicates, that other degradation mechanisms, such as force reduction, might be of higher importance on the long-term behavior.

9 Future Work

The distinct impact of film development was demonstrated through experimental tests carried out in inert gas N_2 and comparative studies conducted in air (standard atmosphere). Considering AlMgSi, the difference between measured and theoretical expected contact resistance was also present in preliminary studies in N_2. However, the sample size was too small to predict a distinct conclusion. Further experimental studies in this field need to be done. Experimental tests with stationary electrical connections, such as bolted busbar joints, should be conducted within an inert gas atmosphere in order to determine the film's impact on the connection resistance. Copper connections could be used for verification purposes before studying aluminum or coated material systems.

Moreover, the thermal operation of long-term tests with crossed rods and bolted busbar joints placed within an inert gas atmosphere would appropriately supplement the results of this research work. It is expected, that the material yielding and therefore the true contact area's increase would not be diminished by film growth due to the significantly lower oxygen presence. Even if film development occurs, the impact should be significantly lower compared to standard atmosphere.

Since the degradation of aluminum-copper joints cannot be entirely explained, this process needs to be studied in more depth to obtain the true physical cause for the contact resistance degradation. Long-term test on stationary electrical connections and contacts (bolted busbar joints and crossed rods) both, have shown partially unstable performances of nickel coated materials. A holistic approach should further consider what impact different kinds of nickel coating, manufacturing technology, cleaning procedure and operating conditions have on nickel coated material systems to enable a stable long-term operation. Moreover, the detected intermetallic phase in the Cu-Ni-Sn system (Fig. 46) should be considered in future works.

The application of lubrications in the field should be carefully considered. Its application might be not necessary for stationary electrical contacts with a constant force and circular contact points if only oxidation is present. However, these results will change if corrosion occurs. Future research on the long-term behavior of chemical reactions should address the impact of corrosion reactions when film development and force reduction occur simultaneously (Fig. 11).

The film's impact on the current constriction should be investigated for power connections such as bolted busbar joints. Because of its nonuniform stress distribution, the spot distribution may alter within the apparent contact area of bolted busbar joints.

The material data for numerical analysis should be experimentally determined for all studied materials since the straight application of raw data found in the literature results may produce inaccurate results. This would be crucial when determining the material's yielding by applying numerical methods.

10 References

[1] R. Holm and E. Holm, *Electric contacts theory and application*. Berlin; New York: Springer, 1967, ISBN: 978-3-662-06688-1.

[2] J. G. Simmons, "Generalized Formula for the Electric Tunnel Effect between Similar Electrodes Separated by a Thin Insulating Film," *Journal of Applied Physics*, vol. 34, no. 6, p. 1793, 1963, ISSN: 00218979, doi: 10.1063/1.1702682.

[3] P. G. Slade, Ed., *Electrical contacts: principles and applications*, Second edition. Boca Raton: CRC Press, Taylor & Francis Group, 2014, ISBN: 978-1-4398-8130-9.

[4] K. R. Lawless, "The oxidation of metals," *Reports on Progress in Physics*, vol. 37, no. 2, pp. 231–316, Feb. 1974, ISSN: 0034-4885, 1361-6633, doi: 10.1088/0034-4885/37/2/002.

[5] R. Holm, "Über metallische Kontaktwiderstände," *Wissenschaftliche Veröffentlichungen aus den Siemens-Werken*, vol. 42, no. 39, pp. 217–271, Sep. 1929.

[6] J. A. Greenwood, "Constriction resistance and the real area of contact," *British Journal of Applied Physics*, vol. 17, no. 12, pp. 1621–1632, Dec. 1966, ISSN: 0508-3443, doi: 10.1088/0508-3443/17/12/310.

[7] M. Braunovic, V. V. Konchits, and N. K. Myshkin, *Electrical contacts fundamentals, applications and technology*. Boca Raton: CRC Press, 2007, ISBN: 978-0-8493-9108-8.

[8] W. Schatt, *Werkstoffwissenschaft*. Stuttgart: Deutscher Verlag für Grundstoffindustrie, 1996, ISBN: 3-342-00675-7.

[9] R. J. Roark, *Roark's formulas for stress and strain*, 7th ed. New York: McGraw-Hill, 2002, ISBN: 0-07-072542-X.

[10] V. L. Popov, *Contact mechanics and friction: physical principles and applications*, English ed. Heidelberg ; New York: Springer, 2010, ISBN: 978-3-642-10802-0.

[11] R. Bürgel, *Handbuch Hochtemperatur- Werkstofftechnik Grundlagen, Werkstoffbeanspruchungen, Hochtemperaturlegierungen und -beschichtungen*. Wiesbaden: Vieweg + Teubner, 2011, ISBN: 978-3-8348-9907-1.

[12] S. Schlegel, "Langzeitverhalten von Schraubenverbindungen mit Stromschienen aus Reinkupfer in der Elektroenergietechnik unter besonderer Berücksichtigung der Temperatur," Dissertation, TU Dresden, Dresden, 2011.

[13] S. Schlegel, S. Grossmann, H. Löbl, M. Hoidis, U. Kaltenborn, and T. Magier, "Joint Resistance of Bolted Copper - Copper Busbar Joints Depending on Joint Force at Temperatures beyond 105 °C," presented at the ICEC, 2010, pp. 1–8, ISBN: 978-1-4244-8174-3, doi: 10.1109/HOLM.2010.5619459.

[14] J. Lienig, "introduction to electromigration-aware physical design," in *Proceedings of the 2006 International Symposium on Physical Design*, San Jose, 2006, p. 39, ISBN: 1-59593-299-2, doi: 10.1145/1123008.1123017.

[15] J. Lienig, "Electromigration and its impact on physical design in future technologies," in *Proceedings of the 2013 ACM international symposium on International symposium on physical design*, Lake Tahoe, 2013, p. 33, ISBN: 978-1-4503-1954-6, doi: 10.1145/2451916.2451925.

[16] R. D. Naybour and T. Farrell, "Degradation mechanisms of mechanical connectors on aluminium conductors," *Proceedings of the Institution of Electrical Engineers*, vol. 120, no. 2, p. 273, 1973, ISSN: 00203270, doi: 10.1049/piee.1973.0060.

[17] M. Braunovic and N. Aleksandrov, "Intermetallic compounds at aluminum-to-copper and copper-to-tin electrical interfaces," in *Proceedings of the Thirty-Eighth IEEE Holm Conference on Electrical Contacts*, Philadelphia, 1992, pp. 25–34, ISBN: 0-7803-0576-0, doi: 10.1109/HOLM.1992.246938.

[18] C. Ruppert and M. Runde, “Thermally Induced Mechanical Degradation of Contact Spots in Aluminum Interfaces,” *IEEE Transactions on Components and Packaging Technologies*, vol. 29, no. 4, pp. 833–840, Dec. 2006, ISSN: 1521-3331, doi: 10.1109/TCAPT.2006.885964.

[19] M. Runde, E. Hodne, and B. Totdal, “Experimental study of the conducting spots in aluminum contact interfaces,” *IEEE Transactions on Components, Hybrids, and Manufacturing Technology*, vol. 13, no. 4, pp. 1068–1073, Dec. 1990, ISSN: 01486411, doi: 10.1109/33.62550.

[20] M. Runde, E. Hodne, and B. Totdal, “Current-induced aging of contact spots,” in *Proceedings of the Thirty Fifth Meeting of the IEEE Holm Conference on Electrical Contacts*, Chicago, 1989, pp. 213–220, doi: 10.1109/HOLM.1989.77942.

[21] M. Runde, “Mass Transport in Stationary Contact Points,” *IEEE Transactions on Components, Hybrids, and Manufacturing Technology*, vol. 10, no. 1, pp. 89–99, Mar. 1987, ISSN: 0148-6411, doi: 10.1109/TCHMT.1987.1134705.

[22] M. Runde, H. Kongsjorden, J. Kulsetas, and B. Totdal, “Detection of a-Spots in Aluminum Contacts,” *IEEE Transactions on Components, Hybrids, and Manufacturing Technology*, vol. 9, no. 1, pp. 77–85, Mar. 1986, ISSN: 0148-6411, doi: 10.1109/TCHMT.1986.1136622.

[23] J. Aronstein and T. K. Hare, “AC and DC electromigration failure of aluminum contact junctions,” *IEEE Transactions on Components and Packaging Technologies*, vol. 28, no. 4, pp. 701–709, Dec. 2005, ISSN: 1521-3331, doi: 10.1109/TCAPT.2005.859676.

[24] K.-H. Tostmann, *Korrosion Ursachen und Vermeidung*. Weinheim: Wiley, 2001, ISBN: 3-527-30203-4.

[25] R. W. Revie, H. H. Uhlig, and H. H. Uhlig, Eds., *Uhlig's corrosion handbook*, 2nd ed. New York: Wiley, 2000, ISBN: 0-471-15777-5.

[26] A. Rönnquist and H. Fischmeister, “The Oxidation of Copper: A Review of published data,” *Journal of the Institute of Metals*, vol. 89, pp. 65–76, 60 1960.

[27] L. Gmelin, *Gmelin-Handbuch der anorganischen Chemie: Kupfer, Teil B -Lieferung 1: Verbindungen bis Kupfertellurate*, 8. ed., vol. 60. Weinheim: Verlag Chemie, 1958.

[28] L. Gmelin, *Gmelin-Handbuch der anorganischen Chemie, Kupfer, Teil D: Elektrische Eigenschaften der Kupferoxide*, 8. ed., vol. 60. Weinheim: Verlag Chemie, 1963.

[29] J. Berg, “Kupfer als Katalysator für Partialoxidationen. Untersuchungen mit instationären kinetischen Methoden,” Dissertation, TU Darmstadt, 2009.

[30] Samsonov, *The Oxide Handbook.* Springer Verlag, 2012, ISBN: 978-1-4615-9599-1.

[31] “Deutsches Kupferinstitut - data sheet Cu-ETP.” [Online]. Available: https://www.kupferinstitut.de/fileadmin/user_upload/kupferinstitut.de/de/Documents/Shop/Verlag/Downloads/Werkstoffe/Datenblaetter/Kupfer/Cu-ETP.pdf. [Accessed: 31-Jan-2015].

[32] S. Dreier, N. Lücke, and S. Großmann, “Long Term Behavior of Electrical Contacts with Crossed Rods under Various Environmental Conditions,” in *Proceedings / The 27th International Conference on Electrical Contacts - ICEC 2014: June 22-26, 2014, Dresden, Germany*, Berlin, 2014, ISBN: 978-3-8007-3624-9.

[33] K. Hauffe, *Oxydation von Metallen und Metallegierungen.* Berlin, Heidelberg: Springer, 2014, ISBN: 3-642-49143-X.

[34] A. T. Fromhold and M. H. Anderson, “Oxidation Kinetics of Epitaxial (100) Copper Films at 25 °C and 50 °C,” *Oxidation of Metals*, vol. 62, no. 3–4, pp. 237–272, Oct. 2004, ISSN: 0030-770X, 1573-4889, doi: 10.1007/s11085-004-7810-z.

[35] N. Cabrera and N. F. Mott, “Theory of the oxidation of metals,” *Reports on Progress in Physics*, vol. 12, no. 1, pp. 163–184, Jan. 1949, ISSN: 00344885, doi: 10.1088/0034-4885/12/1/308.

[36] W. J. Moore, Y. Ebisuzaki, and J. A. Sluss, "Exchange and Diffusion of Oxygen in Crystalline Cuprous Oxide," *The Journal of Physical Chemistry*, vol. 62, no. 11, pp. 1438–1441, Nov. 1958, ISSN: 0022-3654, 1541-5740, doi: 10.1021/j150569a022.

[37] M. Sun, M. G. Pecht, M. A. E. Natishan, and R. I. Martens, "Lifetime resistance model of bare metal electrical contacts," *IEEE Transactions on Advanced Packaging*, vol. 22, no. 1, pp. 60–67, Feb. 1999, ISSN: 15213323, doi: 10.1109/6040.746544.

[38] S. Mrowec and A. Stokłosa, "Oxidation of copper at high temperatures," *Oxidation of Metals*, vol. 3, no. 3, pp. 291–311, May 1971, ISSN: 0030-770X, 1573-4889, doi: 10.1007/BF00603530.

[39] W. H. J. Vernon, "Second experimental report to the Atmospheric Corrosion Research Committee (British Non-Ferrous Metals Research Association)," *Transactions of the Faraday Society*, vol. 23, p. 113, 1927, ISSN: 0014-7672, doi: 10.1039/tf9272300113.

[40] W. H. J. Vernon, "First (experimental) report to the Atmospheric Corrosion Research Committee (of the British Non-Ferrous Metals Research Association)," *Transactions of the Faraday Society*, vol. 19, no. March, p. 839, 1924, ISSN: 0014-7672, doi: 10.1039/tf9241900839.

[41] W. E. Campbell and U. B. Thomas, "The Oxidation of Metals," *Transactions of The Electrochemical Society*, vol. 91, no. 1, p. 623, 1947, ISSN: 00964743, doi: 10.1149/1.3071795.

[42] D. Altenpohl, *Aluminium und Aluminiumlegierungen*. Berlin; New York: Springer, 1965, ISBN: 978-3-662-30245-3.

[43] C. Kammer, *Aluminium-Taschenbuch. 1: Grundlagen und Werkstoffe*, 16. Aufl. Düsseldorf: Aluminium-Verlag, 2002, ISBN: 978-3-87017-274-9.

[44] L. Gmelin, *Gmelin-Handbuch der anorganischen Chemie: Aluminium, Teil B: Die Verbindungen des Aluminiums*, 8th ed., vol. 35. Berlin: Verlag Chemie, 1934.

[45] M. S. Hunter and P. Fowle, "Natural and Thermally Formed Oxide Films on Aluminum," *Journal of The Electrochemical Society*, vol. 103, no. 9, p. 482, 1956, ISSN: 00134651, doi: 10.1149/1.2430389.

[46] K. Thomas and M. W. Roberts, "Direct Observation in the Electron Microscope of Oxide Layers on Aluminum," *Journal of Applied Physics*, vol. 32, no. 1, p. 70, 1961, ISSN: 00218979, doi: 10.1063/1.1735963.

[47] R. C. Evans and J. Pickardt, *Einführung in die Kristallchemie*. Berlin: De Gruyter, 1976, ISBN: 3-11-003976-1.

[48] W. D. Treadwell and A. Obrist, "Über die Bestimmung und Bildung von oxydischen Deckschichten auf Aluminium," *Helvetica Chimica Acta*, vol. 26, no. 6, pp. 1816–1828, Oct. 1943, ISSN: 0018-019X, 1522-2675, doi: 10.1002/hlca.19430260611.

[49] R. Grauer and P. Schmoker, "Die Oxidation von Aluminium-Magnesium-Legierungen in Sauerstoff bei erhöhter Temperatur," *Materials and Corrosion/Werkstoffe und Korrosion*, vol. 27, no. 11, pp. 769–774, Nov. 1976, ISSN: 0947-5117, 1521-4176, doi: 10.1002/maco.19760271102.

[50] J. Zähr, "Einfluss des Oberflächenzustandes auf das thermische Fügen von Aluminium," Dissertation, TU Dresden, Dresden, 2011.

[51] A. Nylund, K. Mizuno, and I. Olefjord, "Influence of Mg and Si on the Oxidation of Aluminum," *Oxidation of Metals*, vol. 50, no. 3–4, pp. 309–325, 1998.

[52] O. Madelung and M. Schulz, Eds., "Silver sulfide (Ag2S) crystal structure," in *Non-Tetrahedrally Bonded Elements and Binary Compounds I*, vol. 41C, Berlin/Heidelberg: Springer-Verlag, 1998, pp. 1–4, ISBN: 3-540-64583-7.

[53] L. Gmelin, *Gmelins Handbuch der anorganischen Chemie Teil B*. Berlin: Verlag Chemie, 1996, ISBN: 0-387-93729-3.

[54] A. K. Rudolphi, C. Bjorkman, T. Imrell, and S. Jacobson, "Conduction through corrosion films on silver plated copper in power contacts," in *Proceedings of the Forty-First IEEE Holm Conference on Electrical Contacts*, Montreal, 1995, pp. 124–134, ISBN: 0-7803-2728-4, doi: 10.1109/HOLM.1995.482865.

[55] "Acanthite Mineral Data." [Online]. Available: http://webmineral.com/data/Acanthite.shtml#.VMifmC4nieg. [Accessed: 28-Jan-2015].

[56] D. K. Burge, J. M. Bennett, R. L. Peck, and H. E. Bennett, "Growth of surface films on silver," *Surface Science*, vol. 16, pp. 303–320, Aug. 1969, ISSN: 00396028, doi: 10.1016/0039-6028(69)90026-0.

[57] L. Gmelin, *Gmelin-Handbuch der anorganischen Chemie: Zinn, Teil C 1: Verbindungen mit Wasserstoff, Sauerstoff, Stickstoff und Halogenen*, vol. 46. Leipzig: Verlag Chemie, 1972, ISBN: 3-527-84610-7.

[58] W. E. Boggs, R. H. Kachik, and G. E. Pellissier, "The Oxidation of Tin," *Journal of The Electrochemical Society*, vol. 108, no. 1, p. 6, 1961, ISSN: 00134651, doi: 10.1149/1.2428013.

[59] W. E. Boggs, P. S. Trozzo, and G. E. Pellissier, "The Oxidation of Tin," *Journal of The Electrochemical Society*, vol. 108, no. 1, p. 13, 1961, ISSN: 00134651, doi: 10.1149/1.2428003.

[60] W. E. Boggs, "The Oxidation of Tin," *Journal of The Electrochemical Society*, vol. 108, no. 2, p. 124, 1961, ISSN: 00134651, doi: 10.1149/1.2428026.

[61] W. K. Choi, "Oxidation process from SnO to SnO2," *Journal of Materials Science Letters*, vol. 16, no. 19, pp. 1551–1554.

[62] E. Leja, J. Korecki, K. Krop, and K. Toll, "Phase composition of SnOx thin films obtained by reactive d.c. sputtering," *Thin Solid Films*, vol. 59, no. 2, pp. 147–155, May 1979, ISSN: 00406090, doi: 10.1016/0040-6090(79)90288-8.

[63] "Romarchite Mineral Data." [Online]. Available: http://webmineral.com/data/Romarchite.shtml#.VMprWy4nieg. [Accessed: 29-Jan-2015].

[64] L. Kenworthy, "The atmospheric corrosion and tarnishing of tin," *Transactions of the Faraday Society*, vol. 31, p. 1331, 1935, ISSN: 0014-7672, doi: 10.1039/tf9353101331.

[65] "Cassiterite Mineral Data." [Online]. Available: http://webmineral.com/data/Cassiterite.shtml#.VeFhuZfw-eg. [Accessed: 29-Aug-2015].

[66] R. Holm, *Die technische Physik der elektrischen Kontakte*. Berlin: Springer, 1941.

[67] L. Gmelin, *Gmelins Handbuch der anorganischen Chemie System-Nr. 57. Teil A, Lieferung 1*. Weinheim: Verlag Chemie, 1967, ISBN: 3-540-93137-6.

[68] M. Antler, W. Abbott, W. Campbell, S. Garte, M. Mamrick, A. Snowdon, and M. Trzeciak, "The Effect of Environment on Electrical Contacts: A Discussion at the 1973 Holm Seminar," *IEEE Transactions on Parts, Hybrids, and Packaging*, vol. 11, no. 1, pp. 57–67, Mar. 1975, ISSN: 0361-1000, doi: 10.1109/TPHP.1975.1135040.

[69] W. H. Abbott, "The development and performance characteristics of mixed flowing gas test environment," *IEEE Transactions on Components, Hybrids, and Manufacturing Technology*, vol. 11, no. 1, pp. 22–35, Mar. 1988, ISSN: 01486411, doi: 10.1109/33.2959.

[70] W. J. Sanders, "Laboratory and Field Evaluation of Connectors and Other Accessories for Aluminum Conductors in Severe Marine Environments," *Transactions of the American Institute of Electrical Engineers. Part III: Power Apparatus and Systems*, vol. 78, no. 4, pp. 1342–1350, Dec. 1959, ISSN: 0097-2460, doi: 10.1109/AIEEPAS.1959.4500552.

[71] W. F. Bonwitt, "An Experimental Investigation of the Electrical Performance of Bolted Aluminum-to-Copper Connections," *Transactions of the American Institute of Electrical Engineers*, vol. 67, no. 2, pp. 1208–1219, Jan. 1948, ISSN: 0096-3860, doi: 10.1109/T-AIEE.1948.5059803.

[72] W. H. Abbott, "The development and performance characteristics of mixed flowing gas test environment," *IEEE Transactions on Components, Hybrids, and Manufacturing Technology*, vol. 11, no. 1, pp. 22–35, Mar. 1988, ISSN: 01486411, doi: 10.1109/33.2959.

[73] W. H. Abbott, "Materials, environment, motion, and electrical contact failure mechanisms," in *Proceedings of the Thirty Fifth Meeting of the IEEE Holm Conference on Electrical Contacts*, Chicago, 1989, pp. 3–11, doi: 10.1109/HOLM.1989.77913.

[74] M. Antler, "Field Studies of Contact Materials: Contact Resistance Behavior of Some Base and Noble Metals," *IEEE Transactions on Components, Hybrids, and Manufacturing Technology*, vol. 5, no. 3, pp. 301–307, Sep. 1982, ISSN: 0148-6411, doi: 10.1109/TCHMT.1982.1135976.

[75] S. Grossmann, "Zur Gestaltung und zum Betriebsverhalten von Steckverbindungen der Elektroenergietechnik," Dissertation, TU Dresden, Dresden, 1988.

[76] H. Böhme and H. Löbl, "Zur Theorie des Lanzeitverhaltens von Aluminium-Schraubverbindungen," *ELEKTRIE*, vol. 41, no. 5, pp. 179–183, 1987.

[77] S. Dreier and S. Grossmann, "Impact of various harsh environmental conditions on the long term behavior of electrical joints," in *26th International Conference on Electrical Contacts*, Beijing, 2012, pp. 502–508, doi: 10.1049/cp.2012.0706.

[78] H. Böhme, *Mittelspannungstechnik: Schaltanlagen berechnen und entwerfen*, 2nd ed. Berlin: Huss-Medien, Verlag Technik, 2005, ISBN: 3-341-01495-0.

[79] R. Bergmann, *Zum Langzeitverhalten des Widerstands elektrischer Stromschienenverbindungen*. Dissertation, TU Dresden, Düsseldorf: VDI Verlag, 1996, ISBN: 3-18-319521-6.

[80] B. W. Callen, B. Johnson, P. King, R. S. Timsit, and W. H. Abbott, "Environmental degradation of utility power connectors in a harsh environment," *IEEE Transactions on Components and Packaging Technologies*, vol. 23, no. 2, pp. 261–270, Jun. 2000, ISSN: 1521-3331, doi: 10.1109/6144.846763.

[81] B. Johnson and M. Braunovic, "Performance of utility power connectors in a saline environment," in *2001 IEEE/PES Transmission and Distribution Conference and Exposition*, Atlanta, 2001, pp. 781–786, ISBN: 0-7803-7285-9, doi: 10.1109/TDC.2001.971337.

[82] E. Takano, "Thin Tunnel Film in Contact Boundary and the Lifetime for Closed Contacts Deteriorated by Ingress of Reactants," in *23rd International Conference on Electrical Contacts*, Sendai, 2006, pp. 51–56.

[83] R. Schneider, "Langzeitverhalten geschraubter Stromschienenverbindungen in der Elektroenergietechnik," Dissertation, TU Dresden, Berlin, 2013.

[84] R. F. Frank and C. P. Morton, "Comparative Corrosion and Current Burst Testing of Copper and Aluminum Electrical Power Connectors," *IEEE Transactions on Industry Applications*, vol. 43, no. 2, pp. 462–468, Mar. 2007, ISSN: 0093-9994, doi: 10.1109/TIA.2006.889973.

[85] T. Tamai, "Effect of humidity on growth of oxide film on surface of copper contacts," in *23rd International Conference on Electrical Contacts*, Sendai, 2006.

[86] S. Pfeifer, "Einfluss intermetallischer Phasen der Systeme Al-Cu und Al-Ag auf den Widerstand stromtragender Verbindungen im Temperaturbereich von 90°C-200°C," Dissertation, TU Dresden, Dresden, 2015.

[87] T. Fuhrmann, S. Schlegel, S. Grossmann, and M. Hoidis, "Investigations on stationary electrical joints with a bare and a silver or nickel coated contact partner regarding the

permissible temperature limit according to ANSI IEEE and IEO," in *2014 IEEE 60th Holm Conference on Electrical Contacts*, New Orleans, 2014, pp. 1–8, ISBN: 978-1-4799-6068-2, doi: 10.1109/HOLM.2014.7031050.

[88] R. Bergmann, H. Löbl, H. Bohme, and S. Grossmann, "Model to assess the reliability of electrical joints," in *18th International Conference on Electrical Contacts*, 1996, pp. 180–188, ISBN: 978-0-7803-3578-3, doi: 10.1109/HOLM.1996.557195.

[89] S. Schoft, *Langzeitverhalten elektrotechnischer Verbindungen unter Berücksichtigung des Kriechens der Leitermaterialien*. Dissertation, Düsseldorf: VDI-Verl, 2008, ISBN: 978-3-18-338121-0.

[90] T. Fuhrmann, S. Schlegel, S. Großmann, and M. Hoidis, "Comparison between nickel and silver as coating materials of conductors made of copper or aluminum used in electric power engineering," in *Proceedings / The 27th International Conference on Electrical Contacts - ICEC 2014: June 22-26, 2014, Dresden, Germany*, Berlin, 2014, ISBN: 978-3-8007-3624-9.

[91] T. Imrell, "The importance of the thickness of silver coating in the corrosion behaviour of copper contacts," in *Proceedings of the Thirty-Seventh IEEE Holm Conference on Electrical Contacts*, Chicago, 1991, pp. 237–243, ISBN: 0-7803-0231-1, doi: 10.1109/HOLM.1991.170829.

[92] Y. Watanabe, Y. Fujinaga, and H. Iwasaki, "Lattice modulation in the long-period superstructure of Cu_3Sn," *Acta Crystallographica Section B Structural Science*, vol. 39, no. 3, pp. 306–311, Jun. 1983, ISSN: 0108-7681, doi: 10.1107/S0108768183002451.

[93] A.-K. Larsson, L. Stenberg, and S. Lidin, "The superstructure of domain-twinned η'-Cu_6Sn_5," *Acta Crystallographica Section B Structural Science*, vol. 50, no. 6, pp. 636–643, Dec. 1994, ISSN: 0108-7681, doi: 10.1107/S0108768194004052.

[94] T. Sasaki, M. Tanaka, and Y. Ohno, "Intermetallic compound formation between lead-free solders (Sn) and Cu or Ni electrodes," *Materials Letters*, vol. 61, no. 10, pp. 2093–2095, Apr. 2007, ISSN: 0167577X, doi: 10.1016/j.matlet.2006.08.021.

[95] P. G. Harris and K. S. Chaggar, "The role of intermetallic compounds in lead-free soldering," *Soldering & Surface Mount Technology*, vol. 10, no. 3, pp. 38–52, Dec. 1998, ISSN: 0954-0911, doi: 10.1108/09540919810237110.

[96] M. Braunovic, "Effect of intermetallic phases on the performance of tin-plated copper connections and conductors," in *Proceedings of the Forty-Ninth IEEE Holm Conference on Electrical Contacts*, 2003, pp. 124–131, ISBN: 0-7803-7862-8, doi: 10.1109/HOLM.2003.1246487.

[97] B. Chao, S.-H. Chae, X. Zhang, K.-H. Lu, J. Im, and P. S. Ho, "Investigation of diffusion and electromigration parameters for Cu–Sn intermetallic compounds in Pb-free solders using simulated annealing," *Acta Materialia*, vol. 55, no. 8, pp. 2805–2814, May 2007, ISSN: 13596454, doi: 10.1016/j.actamat.2006.12.019.

[98] T. Hetschel, *Alterungsverhalten und Lötbarkeit der chemisch abgeschiedenen Zinn-Leiterplattenendoberfläche*. Dissertation, Templin, 2010, ISBN: 978-3-934142-35-0.

[99] J. R. Davis, Ed., *Copper and copper alloys*. Materials Park: ASM International, 2001, ISBN: 0-87170-726-8.

[100] E. Vinaricky and A. Keil, Eds., *Elektrische Kontakte, Werkstoffe und Anwendungen: Grundlagen, Technologien, Prüfverfahren*, 2nd ed. Berlin: Springer, 2002, ISBN: 978-3-540-42431-4.

[101] J. Datta, *Aluminium-Werkstoff-Datenblätter*. Düsseldorf: Aluminium-Verlag, 2003, ISBN: 3-87017-281-9.

[102] D. Mende and G. Simon, *Physik: Gleichungen und Tabellen*, 16th ed. München: Carl Hanser Verlag, 2013, ISBN: 978-3-446-43754-8.

[103] K.-H. Grote and J. Feldhusen, Eds., *Dubbel*. Berlin, Heidelberg: Springer Berlin Heidelberg, 2014, ISBN: 978-3-642-38890-3.

[104] D. R. Lide and Chemical Rubber Company, Eds., *CRC Handbook of Chemistry and Physics, Internet Version 2005*. Boca Raton, FLa.: CRC Press, 2005.

[105] H. Gremmel, ABB-Calor-Emag-Schaltanlagen AG, and ABB-Calor-Emag-Mittelspannung GmbH, Eds., *Schaltanlagen*, 10th ed. Berlin: Cornelsen, 1999, ISBN: 978-3-464-48235-3.

[106] H. Mang, G. Hofstetter, and J. Eberhardsteiner, *Festigkeitslehre*, 3., aktualisierte Aufl. Wien: Springer, 2008, ISBN: 978-3-211-72453-8.

[107] C. Hildmann, "Study on the Electrical and Mechanical Behavior of Power Connections for Overhead Power Lines," TU Dresden, unpublished work.

[108] R. P. Carreker, "Tensile Deformation of Silver as a Function of Temperature, Strain Rate, and Grain Size," *Journal of Metals*, pp. 112–115, Jan. 1957.

[109] M. Hasegawa and H. Fukutomi, "Microstructural Study on Dynamic Recrystallization and Texture Formation in Pure Nickel.," *Material Transactions*, vol. 43, no. 5, pp. 1183–1190, 2002, ISSN: 1345-9678, doi: 10.2320/matertrans.43.1183.

[110] S. Hotta, K. Matsumoto, T. Murakami, T. Narushima, and C. Ouchi, "Dynamic and Static Restoration Behaviors of Pure Lead and Tin in the Ambient Temperature Range," *MATERIALS TRANSACTIONS*, vol. 48, no. 10, pp. 2665–2673, 2007, ISSN: 1347-5320, 1345-9678, doi: 10.2320/matertrans.MRA2007078.

[111] A. Erk and M. Schmelzle, *Grundlagen der Schaltgerätetechnik; Kontaktglieder und Löscheinrichtungen elektrischer Schaltgeräte der Energietechnik*. Berlin, New York: Springer-Verlag, 1974, ISBN: 978-3-540-06075-8.

[112] L. Kogut and I. Etsion, "Elastic-Plastic Contact Analysis of a Sphere and a Rigid Flat," *Journal of Applied Mechanics*, vol. 69, no. 5, p. 657, 2002, ISSN: 00218936, doi: 10.1115/1.1490373.

[113] R. L. Jackson and I. Green, "A Finite Element Study of Elasto-Plastic Hemispherical Contact Against a Rigid Flat," *Journal of Tribology*, vol. 127, no. 2, p. 343, 2005, ISSN: 07424787, doi: 10.1115/1.1866166.

[114] R. Jackson, I. Chusoipin, and I. Green, "A Finite Element Study of the Residual Stress and Deformation in Hemispherical Contacts," *Journal of Tribology*, vol. 127, no. 3, p. 484, 2005, ISSN: 07424787, doi: 10.1115/1.1843166.

[115] D. Mercier, V. Mandrillon, A. Holtz, F. Volpi, M. Verdier, and Y. Bréchet, "Quantitative Evolution of Electrical Contact Resistance between Aluminum Thin Films," in *2012 IEEE 58th Holm Conference on Electrical Contacts*, Portland, 2012, pp. 1–8, ISBN: 978-1-4673-0781-9, doi: 10.1109/HOLM.2012.6336550.

[116] R. Bergmann, H. Löbl, H. Böhme, and S. Großmann, "Model to describe the chemical ageing behaviour of electrical busbar joints," *European Transactions on Electrical Power*, vol. 7, no. 5, pp. 345–350, Sep. 2007, ISSN: 1430144X, 15463109, doi: 10.1002/etep.4450070508.

[117] S. Schlegel, S. Grossmann, M. Lakner, and T. Schoenemann, "Investigations on material structural changes on electrical joints at high contact temperature," in *26th International Conference on Electrical Contacts*, Beijing, 2012, pp. 7–11, ISBN: 978-1-84919-508-9, doi: 10.1049/cp.2012.0613.

[118] F. Blumenroth, *Zum Langzeitverhalten von Steckverbindungen mit Schraubenfedern in Anlagen der Elektroenergietechnik*. Dissertation, Aachen: Shaker Verlag, 2010, ISBN: 978-3-8322-9506-6.

[119] M. Braunovic, V. V. Izmailov, and M. V. Novoselova, "A model for life time evaluation of closed electrical contacts," in *Proceedings of the Fifty-First IEEE Holm Conference on Electrical Contacts*, Chicago, 2005, pp. 217–223, ISBN: 978-0-7803-9113-0, doi: 10.1109/HOLM.2005.1518247.

[120] V. V. Izmailov and N. N. Dzektser, "On the Lifetime of Stationary Contacts," in *16th International Conference on Electrical Contacts*, Loughborough, 1992.

[121] R. D. Malucci, "Comparison of Model Predictions of Current Density Variation across Simulated Contact Regions," in *Proceedings of the 56th IEEE Holm Conference on Electrical Contacts*, Charleston, 2010, pp. 1–9, ISBN: 978-1-4244-8174-3, doi: 10.1109/HOLM.2010.5619536.

[122] R. D. Malucci, "The Effects of Current Density Variations in Power Contact Interfaces," in *2011 IEEE 57th Holm Conference on Electrical Contacts*, Minneapolis, 2011, pp. 1–7, ISBN: 978-1-61284-650-7, doi: 10.1109/HOLM.2011.6034781.

[123] R. D. Malucci, "Current Redistribution across an Aging Contact Interface," in *2012 IEEE 58th Holm Conference on Electrical Contacts*, Portland, 2012, pp. 1–8, ISBN: 978-1-4673-0781-9, doi: 10.1109/HOLM.2012.6336578.

[124] S. Dreier, S. Großmann, G. Moustafa, D. G. dos Santos, B. R. P. da Rocha, and G. M. Macedo, "Long-term Behavior of Stationary Electrical Connections in Areas with Harsh Natural Environmental Conditions," *Cigre Science & Engineering*, vol. 3, pp. 40–54, Oct. 2015.

[125] Institute of Electrical Power Systems and High-Voltage Engineering, TU Dresden, Unpublished work, 2011.

[126] R. Bergmann, H. Löbl, H. Bohme, and S. Großmann, "Calculation of the lifetime of electrical busbar joints," *European Transactions on Electrical Power*, vol. 7, no. 6, pp. 403–408, Sep. 2007, ISSN: 1430144X, 15463109, doi: 10.1002/etep.4450070606.

[127] S. Pfeifer, S. Grossmann, and S. Schlegel, "Alterung stationärere elektrischer Verbindungen durch Interdiffusion," in *Elektrische und optische Verbindungstechnik 2013*, Lemgo, 2013, pp. 42–53, ISBN: 978-3-00-041043-7.

[128] H. Hoeft, *Elektrische Kontakte / physikalische Grundlagen*. Berlin: Verlag Technik, 1977.

[129] W. Schufft, *Taschenbuch der elektrischen Energietechnik*. München: Fachbuchverlag Leipzig, 2007, ISBN: 978-3-446-40475-5.

[130] J. A. Assimos and D. Trivich, "The photoelectric threshold, work function, and surface barrier potential of single-crystal cuprous oxide," *Physica Status Solidi (a)*, vol. 26, no. 2, pp. 477–488, Dec. 1974, ISSN: 00318965, 1521396X, doi: 10.1002/pssa.2210260210.

Standards

DIN EN 13601: 2013, Copper and copper alloys – Copper rod, bar and wire for general electrical purposes.

DIN EN 573-3: 2009, Aluminium and aluminium alloys – Chemical composition and form of wrought products – Part 3: Chemical composition and form of products.

DIN EN ISO 8044: 1999, Corrosion of metals and alloys - Basic terms and definitions.

DIN EN 60721-1: 1990 + A1: 1992 + A2: 1995, Classification of environmental conditions- Part 1: Environmental parameters and their severities.

DIN EN 60068-1: 1994 + Corrigendum 1988 + A1: 1992, Environmental testing – Part 1: General and guidance.

IEC 62271-1: 2007 + A1: 2011, High-voltage switchgear and controlgear – Part 1; Common specifications.

11 List of Figures

12 List of Tables

13 Appendix

Appendix A

Pilling-Bedworth-Ratio

The Pilling-Bedworth-Ratio (PBR) can be used to assess the self-limiting properties of an oxide film. The PBR is defined as ratio of the oxide and metal unit cell volume, $PBR = V_{oxide} / V_{metal}$. For PBR < 1 tensile stress enlarges with increasing film thickness until the tensile strength is exceeded and the film breaks up. The film is porous and non-passivating. Passivating films can commonly develop for PBR > 1. Compressive stress can develop along the surface if the growth direction is inward. For low stress levels the films remains unimpaired. If the stress levels exceeds the film's strength (PBR >> 1) the film will be lifted and break up. As a result thereof oxygen moves into the cracks of the film towards the metal surface and develop new oxide until a passivating film thickness is reached.

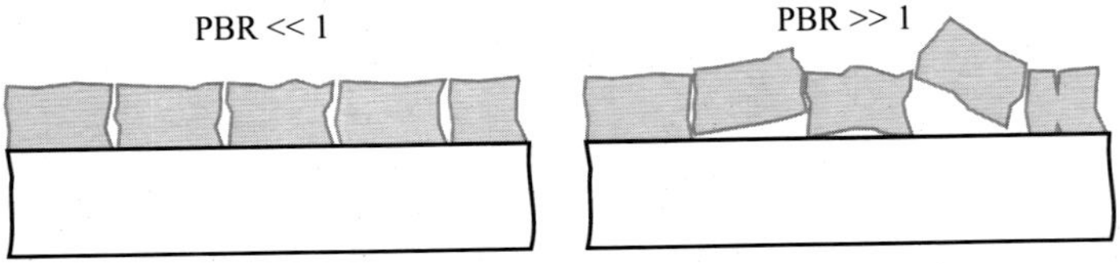

Fig. 73: Films with PBR << 1 show partially cracks but the film remains on the surface, films with PBR >> 1 lift up and fracture [11]

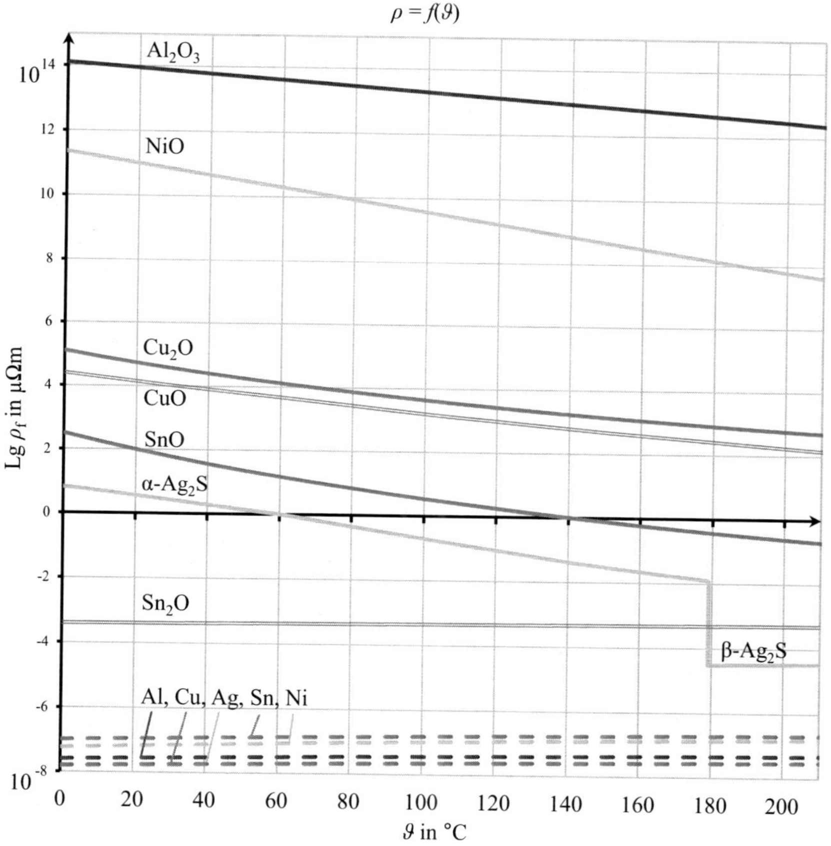

Fig. 74: Resistivity dependent on temperature according to Eq. (25), [30], [53], [57], [62], [100], [data research by Michael Gatzsche]

Appendix B

Chemical Constitution Substrate

The results of the ICP-OES and the TECG analysis have shown that the majority of the elements is below the detection limit and the others are within the defined boundaries according to standards (Tab. 23).

Tab. 23: Results of the chemical analysis with ICP-OES and TECG (measurement done by fem)

	Cu CW004A in %	Set point in % according to DIN EN 13601		Al EN AW-1050A in %	Set point in % according to DIN EN 573-3	AlMgSi EN AW-6060 in %	Set point in % according to DIN EN 573-3
Bi	<0.0002	max. 0.0005	Bi	<0.01		<0.01	
Cr	<0.0001		Cr	<0.01		<0.01	max. 0.05
Cd	<0.0001		Cd	<0.01		<0.01	
Fe	0.0018		Fe	0.16	max. 0.4	0.18	max. 0.3
Mg	<0.0001		Mg	<0.01	max. 0.05	0.4	0.35 - 0.6
Mn	<0.0001		Mn	<0.01	max. 0.05	0.03	max. 0.10
Ni	0.0011		Ni	<0.01		<0.01	
Pb	0.0042	max. 0.0050	Pb	<0.01		<0.01	
Si	<0.001		Si	0.04	max. 0.25	0.472	0.3 - 0.6
Sn	0.0026		Sn	<0.01		<0.01	
Ti	<0.0001		Ti	<0.01	max. 0.05	<0.01	max. 0.1
Zn	0.005		Zn	<0.01	max. 0.07	<0.01	max. 0.15
Zr	<0.0001		Zr	<0.01		<0.01	
Al	<0.0001		Al	Rest		Rest	
Cu	Rest		Cu	<0.01	max. 0.05	<0.01	max. 0.1
Ag	0.0014		V	<0.01		<0.01	
As	0.0002		Ga	0.2		0.01	
Be	<0.0001						
Co	<0.0001						
O	0.021	max. 0.04					
P	<0.0001						
S	0.0011						
Sb	0.0003						
Se	<0.0002						
Te	<0.0002						

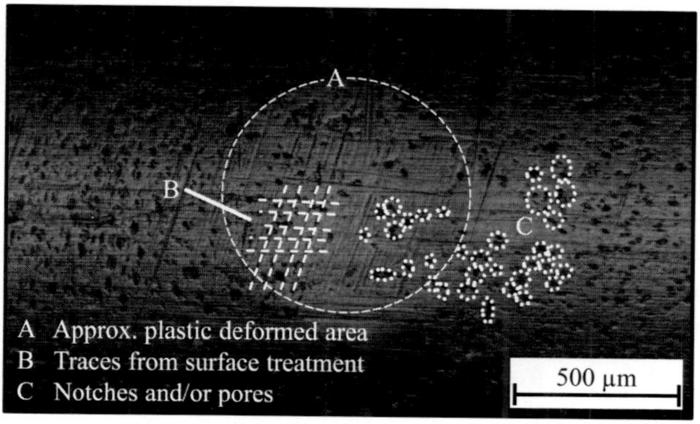

Fig. 75: Surface condition of nickel plated copper rods (Cu-Ni)

Appendix C

Mechanical Deformed Contact Area

A considerable amount of literature has been published on the deformation of the contact area of a contact point. The generalizability of much published research on this issue is problematic. For practicality users tend to use simple approaches easy to handle. On the other hand computed solutions based on the finite element method (FEM) enable higher accuracy and the possibility to link various problems into a multi-physics approach. It was considered that quantitative measures by microscopy would usefully supplement and the qualitative stress analysis by FEM. All three procedures (Fig. 76) have been used to determine the mechanical deformed contact area.

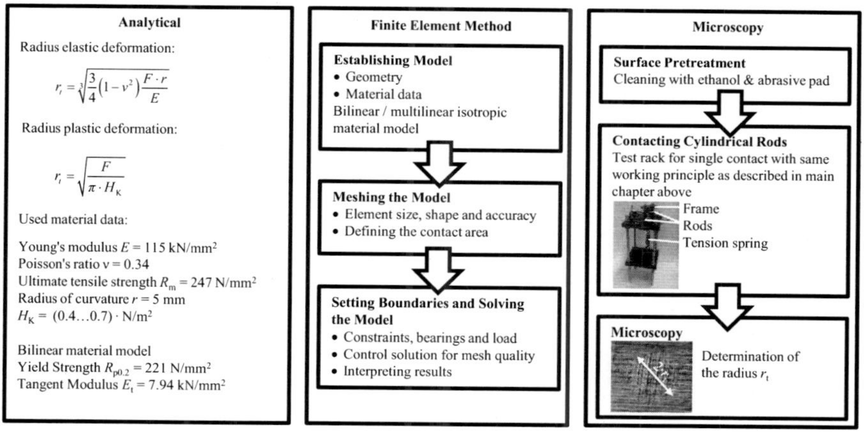

Fig. 76: Methods to determine the mechanical deformed area [12], [32], [99], [128], [129]

Tab. 24: Plastic strain and mechanical stress data

Cu [12]		Ag [108]		Sn [110]		Ni [109]	
Plastic strain	Mechanical Stress in MPa	Plastic strain	Mechanical Stress in MPa	Plastic strain	Mechanical Stress in MPa	Plastic strain	Mechanical Stress in MPa
0.00E+00	211.2	0	46	0	2	0	80
4.98E-05	212.2	0.005	57	0.001	3	0.008	100
1.06E-04	213.5	0.01	65	0.002	6	0.017	120
1.56E-04	214.5	0.02	81	0.003	8	0.024	140
2.09E-04	215.5	0.03	93	0.004	10	0.043	180
2.82E-04	216.8	0.04	104	0.005	12	0.055	200
3.52E-04	218	0.05	114	0.0055	14	0.08	240
4.25E-04	219.2	0.06	123	0.006	14.8	0.11	280
5.02E-04	220.3	0.07	132	0.007	15.3	0.128	300
5.71E-04	221.3	0.08	140	0.008	15.8	0.168	340
6.44E-04	222.3	0.09	148	0.013	17.3	0.18	349
		0.1	156	0.018	18.8	0.19	356
		0.11	162	0.023	19.5	0.2	364
		0.12	169	0.028	20.3	0.223	380
		0.13	175	0.033	21	0.256	400
		0.14	180	0.038	21.5	0.2947	420
		0.15	186	0.048	22.3	0.336	440
		0.16	191	0.058	23	0.361	450
		0.17	197	0.068	23.5	0.3865	460
		0.18	201	0.078	23.8	0.414	470
		0.19	205	0.088	24.1	0.444	480
		0.2	210	0.098	24.3	0.4765	490
		0.21	214	0.148	24.9	0.513	500
		0.217	216	0.198	25.3	0.553	510
		0.23	220	0.248	25.5	0.594	520
		0.24	224	0.298	25.6	0.642	530
		0.25	227	0.348	25.6	0.69	540
		0.26	230	0.398	25.7	0.737	550
		0.27	232	0.448	25.8	0.789	560
		0.28	235	0.498	26.1	0.835	570
		0.287	237	0.548	26.4	0.886	580
						0.91	585

Appendix D

Impact of Films on the Aging of Circular Multiple-Spot Contacts

The here considered film formation takes place in radial direction at the contact sports. The considered contact spots consist of a circular cluster with cluster radius α_c and a number n of circular spots with radius a (Fig. 7). For the film growth three different procedures seems to be possible (Fig. 77).

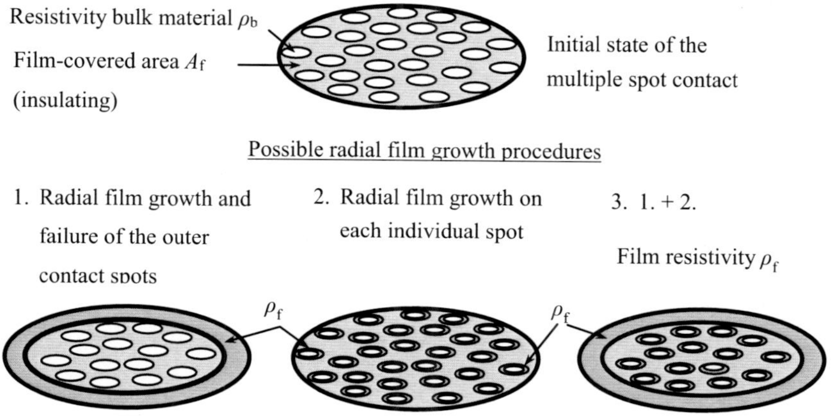

Fig. 77: Principal procedures of possible film growth at multiple spot contacts

The procedures in Fig. 77 were applied to multiple spot models for the calculation of the contact resistance. For n equal circular contact spots with radius a and nearly equal distance between the center of the spots, the calculation of the contact resistance can be done with the approach of Holm ([5], Eq. (17)) and the approach of Greenwood ([6], Eq. (19)). In Holm's approach the influence and the interaction of all spots is given by cluster radius α_c and in Greenwood's approach by the number of the spots and their distance between the centers of the spots. The term ρ / (2 n a) consider the individual influence of the a-spots on the contact resistance. In [1] Holms states another approach for the constriction resistance of n equal hemi-ellipsoids in contact with a plane surface when no film is present. Holm assumes the ellipsoid parameters $\alpha = \beta = a$ and a plane surface parallel to the base area of the hemi-ellipsoids and a distance $2l$ between the a-spot centers. For two contact members with the same hemi-ellipsoid surface the following applies [78]:

$$R(n,a,l) = \frac{\rho}{\pi \cdot n \cdot a} \arctan \frac{\sqrt{l^2 - a^2}}{a} - 1{,}2\rho \frac{\sqrt{l^2 - a^2}}{A_\mathrm{a}} + \frac{\rho}{2\alpha_\mathrm{c}} \tag{41}$$

The apparent contact area A_a and the cluster radius α_c depend on the shape and the dimension of the contact members. Both can differ for other shapes. Because of the large plastic deformed area as well as the apparent contact area and large number n for stationary electrical contacts and connections applied in power engineering, it can be seen, that the results obtained by Eq. (41) would be similar to Eq. (17).

$$R = R_{ca} + R_{\alpha} = \frac{\rho}{2na} + \frac{\rho}{2\alpha_c} \quad (17) \qquad R = \frac{\rho}{2na} + \frac{\rho}{\pi n^2} \sum_{i \neq j} \sum \frac{1}{s_{ij}} \quad (19)$$

In the subchapters below Eqs. (17) and (19) are considered for different proceedings of contact degradation by radial film growth on multi spots and compared among each other.

Radial Film Growth and Failure of the Outer Contact Spots

The radial film growth at the periphery of a multi-spot contact would result in a reduction of the cluster radius α_c. Further the outer spots will fail and the spot number n reduces. The radius a will remain steady. As result, the current would redistribute and the current per spot would increase. The contact resistance would increase.

For a homogenous distribution of individual spots within the cluster, their number n pots results from the ratio of the cluster area and a quadratic area around each spot. The distance between each spot center was defined as 2 l.

$$n = \frac{\pi \cdot \alpha^2}{(2l)^2} \quad (42)$$

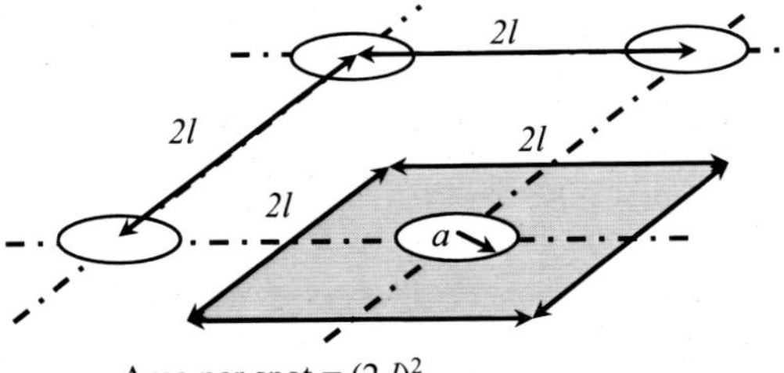

Fig. 78: Quadratic structure of spots with exemplary corresponding area for one spot

With Eq.(42), ρ / (2 n a) can be now defined as:

$$\frac{\rho}{2n \cdot a} \rightarrow \frac{2\rho \cdot l^2}{\pi \cdot \alpha^2 \cdot a} \quad (43)$$

The ratio l/a can be used to determine distance between the spots as well as the interdependence of the individual spots. According to measurements on stationary electrical contacts and connections, Böhme [78] determines an noticeable influence between the spots at $l/a < 10$.

The result of the calculation of Eqs. (17) and (19) for procedure one of Fig. 39, the radial film growth and failure of the outer contact spots can be seen below:

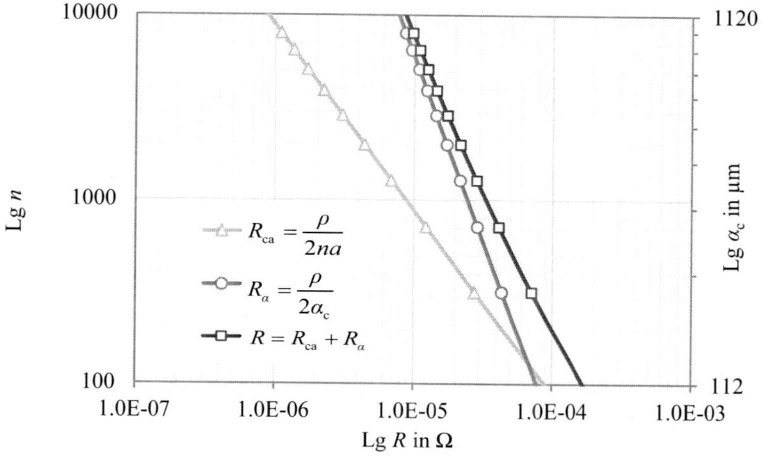

Fig. 79: Contact resistance R, cluster resistance R_α and constriction resistance R_{ca} of n parallel spots dependent on cluster radius α with $a = 1$ µm, $l/a = 10$ and resistivity ρ_{Cu}

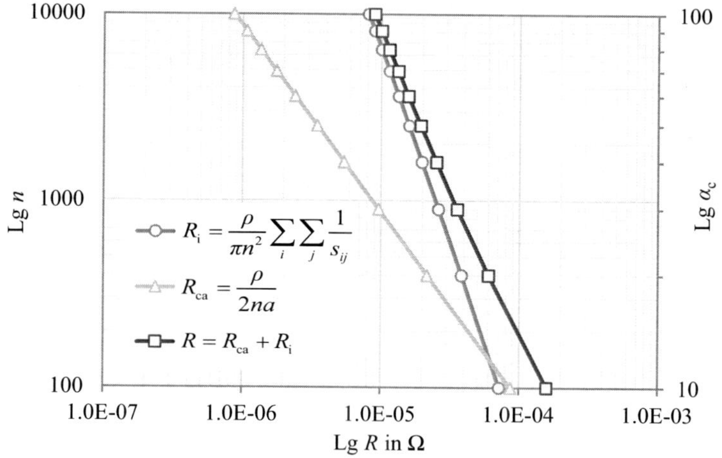

Fig. 80: Contact resistance R, interaction resistance R_i and constriction resistance R_{ca} of n parallel spots dependent on the cluster size $\sqrt{n}$ of a quadratic cluster matrix ($\sqrt{n}$, $\sqrt{n}$) with $a = 1$ µm, $l/a = 10$ and resistivity ρ_{Cu}

In Fig. 79, with decreasing cluster radius and therefore decreasing spot number the influence of the remaining spots on the contact resistance increases. The contact resistance in Fig. 80 decreases with smaller cluster size and therefore also with decreasing spot number. For constant spot size and distance, the influence of the constriction resistance of the individual spots R_{ca}

increases. It is obvious from both figures that the models have similar behavior for decreasing spot number for the considered multi spot contact.

Radial Film Growth on each Individual Spot

The result of the calculations of Eqs. (17) and (19) for proceeding 2, the radial film growth on each individual spot, can be seen below.

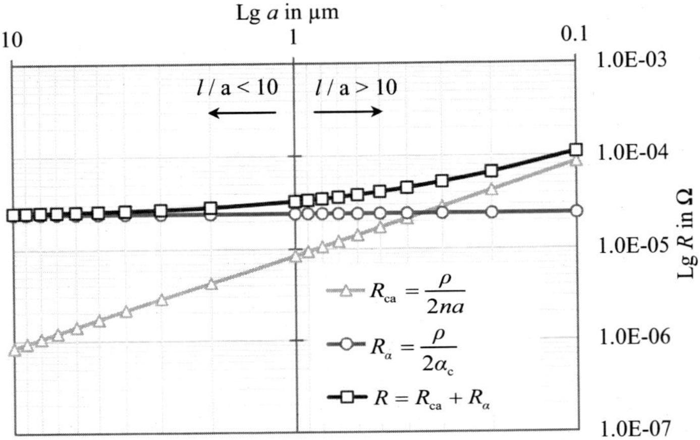

Fig. 81: Contact resistance R, cluster resistance R_{α} and constriction resistance R_{ca} dependent on decreasing spot radius a with $\alpha = 361$ µm, $n = 1024$ and resistivity ρ_{Cu}

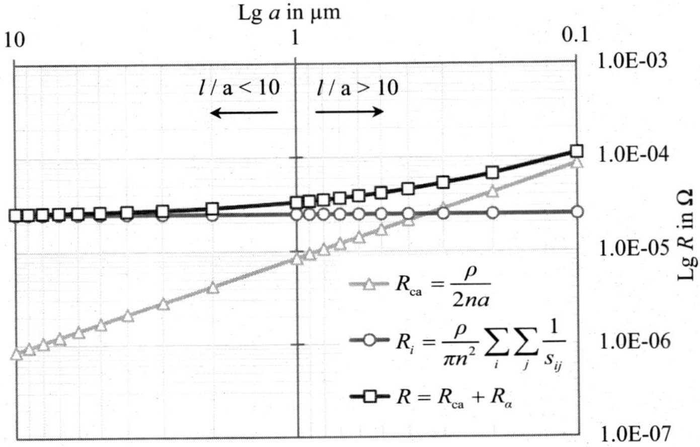

Fig. 82: Contact resistance, interaction resistance R_i and constriction resistance R_{ca} dependent on decreasing spot radius a with $\alpha = 361$ µm cluster size $\sqrt{n} = 32$ and resistivity ρ_{Cu}

It can be seen from Fig. 81 and Fig. 82 that the constriction resistance for n parallel spots will determine the contact resistance for small spot radii a. However, it has to be mentioned that if the mean free path of the electrons is larger than the spot size, other effects such as the Sharvin resistance can be dominant and might change the calculated results R_{ca} for small spot radii with a few nm. The interaction term does not significantly change for $l/a << 10$. The spot interaction and the cluster radius determine the contact resistance for large spot radii a. The determined contact resistances in Fig. 81 and Fig. 82 are nearly equal. The reason for this is that the spot distance is defined as the distance between the spot centers. In [6], Greenwood does not give a distinct information how to define the distance between the spots. The distance between the centers would be no longer applicable if the spots would touch each other at the outer periphery.

Radial Film Growth and Failure of the Outer Contact Spots and Radial Film Growth on each Individual Spot

According to procedure one and two in Fig. 77, the radial film growth at the cluster periphery and the radial film growth on each individual spot were superimposed. In Fig. 83, it can be seen from the figure's left part that for small cluster radii and large spot radii, the resistance is determined through the n parallel spots. The larger the spot radius the lower is the influence of the individual spots. For the approach of Greenwood on the right in Fig. 83, it can be seen, that for large spot numbers the resistance is determined by the interaction term. The interaction term of Greenwood's approach and the cluster radius of Holm's approach might be a lower boundary for the contact resistance. Only for small clusters / low spot number and small spot radii the resulting contact resistance is determined by the constriction of the n individual spots.

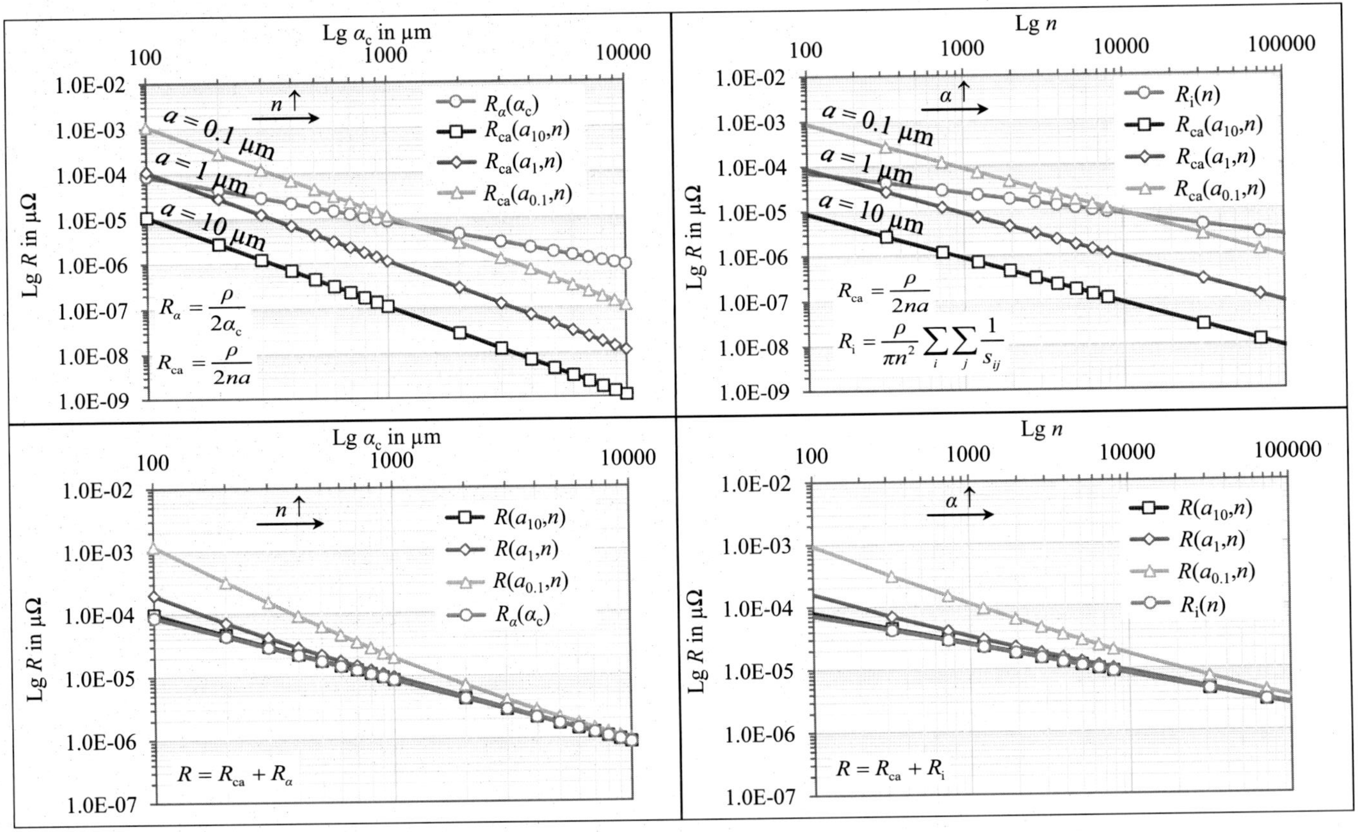

Fig. 83: Resistance dependent R on the cluster radius α_c and the corresponding spot number n for different spot radii a

Exemplary Verification

From the considerations above it is clear that the approach of Holm and Greenwood might result in equal calculated resistances. For one example a verification of the presented approaches with the measured data of Chapter 6.3 was done. The data were gained from measurements (Fig. 27) according to the procedures described in Chapter 6.3 (Fig. 25 and Fig. 26). The microscopic inspection was conducted to determine the radius of the plastic deformed area. An exemplary image is given below:

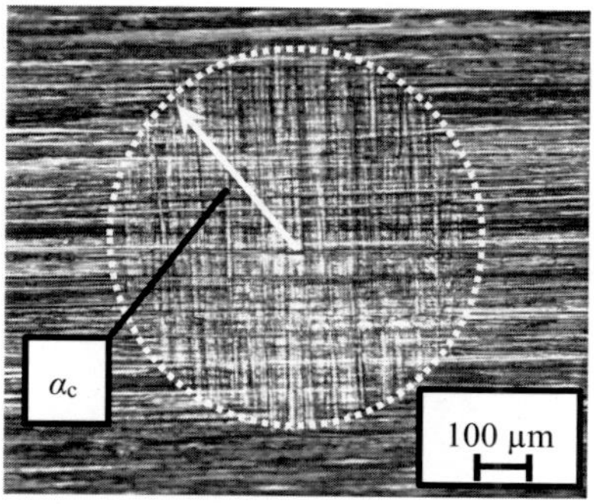

Fig. 84: Contact point image of crossed copper rods with cluster radius $\alpha_c \approx 350$ μm and measured contact resistance R =(26.6…30.9) μΩ

The circular plastic deformed area is easily recognizable. From the given cluster radius the corresponding number of n parallel spots was determined. A homogenous spot distribution with equal distances between the spots was assumed. The spot radius a = (0.1…100) μm and the spot distance $2 \cdot l$ = (1…100) μm were systematically varied. As result the spot number n differs. Cluster radius and measured contact resistance remained constant. In Fig. 83 the contact resistance is determined according to the approaches of Holm [5], and Greenwood [6]. For most of the selected parameter sets in Fig. 83 there is somehow an intersection with the measured resistance. The intersection point is mostly close to the area where half of the spot distance is similar to a. For $l < a$ the spots would touch each other and the calculation is no longer valid. This area (grey shaded area, Fig. 83) is not feasible. In the calculations it was found that for high spot number ($n > 100$) and larger spot radii ($a > 0.5$ μm) the contact resistance is mainly determined by the cluster radius for Holm's approach or the interaction term for Greenwood's approach. These boundaries were exceeded in all measurement in Chapter 6.3. The calculated results of both approaches are basically similar.

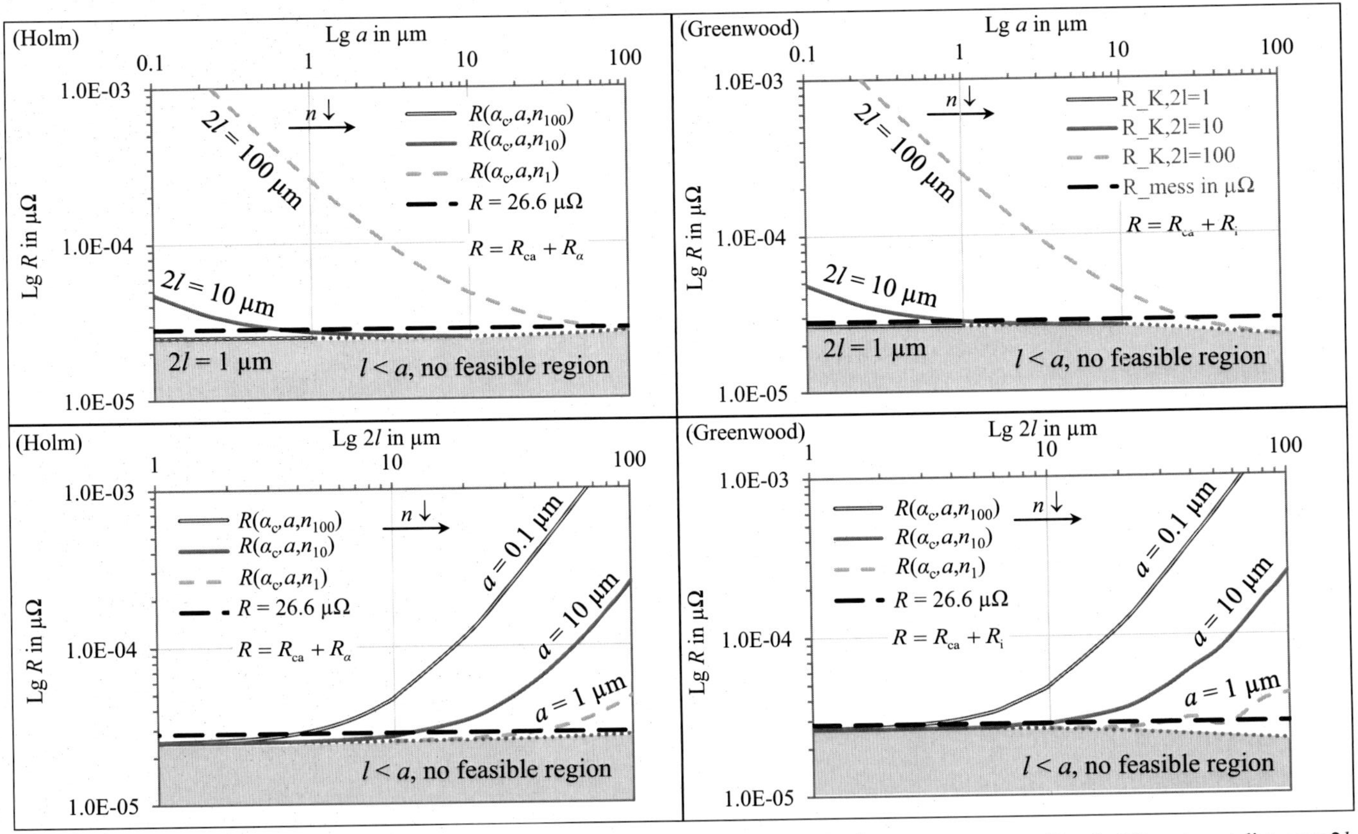

Fig. 85: Contact resistance R dependent on cluster radius α_c & corresponding spot number n for different spot radii a & different spot distances $2l$

Ellipsoid Model for Individual Bulk Material

The application of the ellipsoid model for coatings can be done if $\rho_{coating} \approx \rho_{bulk}$. The solution for a single material is given by Holm [1]:

$$R_{\mu} = \frac{\rho}{4\pi}\int_{0}^{\mu}\frac{d\mu}{\left(a^{2}+\mu\right)\cdot\sqrt{\mu}} = \frac{\rho}{2\pi}\int_{0}^{\sqrt{\mu}}\frac{dz}{\left(a^{2}+z^{2}\right)} = \frac{\rho}{2\pi\cdot a}\arctan\frac{\sqrt{\mu}}{a} \tag{44}$$

Ellipsoid Model for Substrate and Coating with $\rho_{coating} \approx \rho_{bulk}$

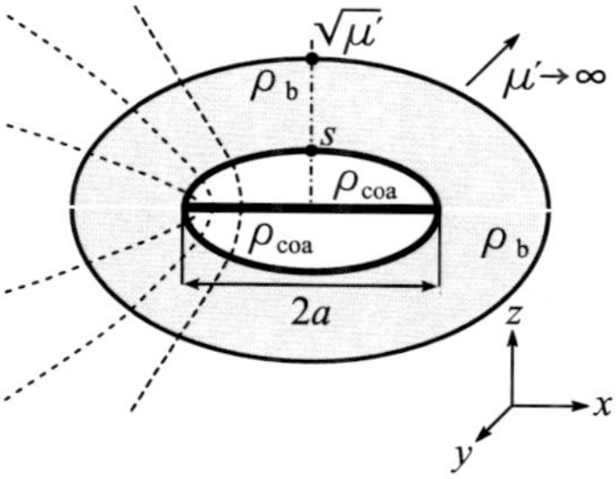

Fig. 86: Profile a-spot with equipotential surfaces, exemplary segment of lines of current flow and marked areas of bulk (grey shaded) and coating for both contact bodies

The current flow in Fig. 86 occurs through the circular contact area (Radius $r = a$) with infinite conductivity in the same way as described in Holm's Ellipsoid model [1]. For coatings with $\rho_{coating} \approx \rho_{bulk}$ the interval of the integral of the Ellipsoid model can be adapted to the new bound of the coating $[0, s^2]$ and the substrate $[s^2, \mu']$. The parameter $\sqrt{\mu'}$ is the height of the corresponding semi-ellipsoid for one contact body, s is the coating thickness and ρ_{coa} & ρ_b are the corresponding resistivity for coating and substrate (bulk). The definite integral can now be solved:

$$R = \frac{\rho_{coa}}{4\pi}\int_{0}^{s^2}\frac{d\mu}{\left(a^{2}+\mu\right)\cdot\sqrt{\mu}} + \frac{\rho_{b}}{4\pi}\int_{s^2}^{\mu'}\frac{d\mu}{\left(a^{2}+\mu\right)\cdot\sqrt{\mu}} \tag{45}$$

$$R = \frac{\rho_{coa}}{4\pi}\cdot\left[\frac{2}{a}\cdot\arctan\frac{\sqrt{\mu}}{a}\right]_{0}^{s^2} + \frac{\rho_{b}}{4\pi}\cdot\left[\frac{2}{a}\cdot\arctan\frac{\sqrt{\mu}}{a}\right]_{s^2}^{\mu'} \tag{46}$$

$$R = \frac{\rho_{coa}}{2\pi\cdot a}\cdot\arctan\frac{s}{a} + \frac{\rho_{b}}{2\pi\cdot a}\cdot\arctan\frac{\sqrt{\mu'}}{a} - \frac{\rho_{b}}{2\pi\cdot a}\cdot\arctan\frac{s}{a} \tag{47}$$

$$R=\frac{\rho_{\text{coa}}}{2\pi\cdot a}\cdot\arctan\frac{s}{a}+\frac{\rho_{\text{b}}}{2\pi\cdot a}\cdot\left(\arctan\frac{\sqrt{\mu'}}{a}-\arctan\frac{s}{a}\right) \tag{48}$$

for $\mu' \rightarrow \infty$:

$$R=\frac{\rho_{\text{coa}}}{2\pi\cdot a}\cdot\arctan\frac{s}{a}+\frac{\rho_{\text{b}}}{2\pi\cdot a}\cdot\left(\frac{\pi}{2}-\arctan\frac{s}{a}\right) \tag{49}$$

For both sides of the contact spot:

$$R=\frac{\rho_{\text{coa}}}{\pi\cdot a}\cdot\arctan\frac{s}{a}+\frac{\rho_{\text{b}}}{\pi\cdot a}\cdot\left(\frac{\pi}{2}-\arctan\frac{s}{a}\right)=\frac{\rho_{\text{b}}}{2\cdot a}+\frac{1}{\pi\cdot a}\cdot\left(\rho_{\text{b}}-\rho_{\text{coa}}\right)\cdot\arctan\frac{s}{a} \tag{50}$$

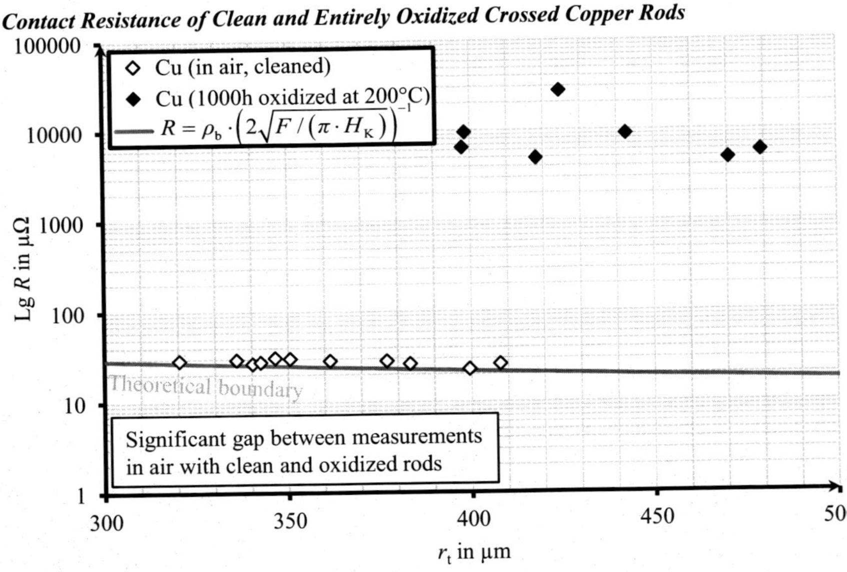

Fig. 87: Contact resistance R in air dependent on the radius of the plastic deformed area r_t plotted for individual measurements of copper rods in air (clean and oxidized) as well as R determined by Eqs. (24) and (33) with the data of Cu from [78] and Tab. 8 and variable force F

Appendix E

Resistance Measurement

During the resistance measurement the test circuits were powered down. The measurements were conducted at ambient temperature. Only for the tests in oven in Chapter 6.5 and 6.7, the resistance measurements was partly conducted during thermal operation. The sample temperature was determined during the measurement process with ±2.5 K accuracy. Therefore, an Ahlborn ALMEMO® 5690 with punched thermocouples (crossed rods) or a temperature sensor ALMEMO® 2590 (bolted busbar joints) was used.

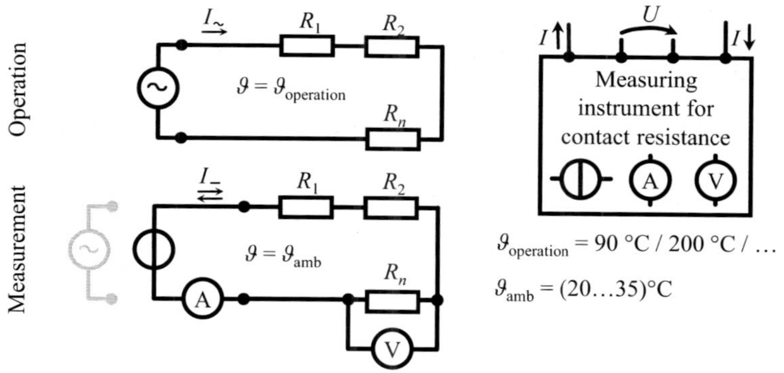

Fig. 88: Schematic circuit for electrical operation and measurement process with ammeter (A), voltmeter (V), voltage sources, samples R_1, R_2…R_n and measuring instrument

The thermoelectric effect was considered by comparative measurements in both directions, either manually or automatically by the used devices.

Tab. 25: Used microohmmeters with corresponding data

Microohmmeter	Voltage type	Current range in A	Measuring range in μΩ	Measurement. error in %
MO2A	DC	2…20	$0.01…2\cdot10^3$	± 0.5
LoRe	DC	≤ 90	$0.01…5\cdot10^5$	≤ 0.8
MI 3252	DC	0.1…100	$1\cdot10^{-3}…2\cdot10^7$	± 0.25

Appendix F

Experimental Studies in Inert Gas Atmosphere

The experimental methodology used in Chapter 6.3 for the contact resistance measurement in N_2 inert gas atmosphere was verified with copper. Because of oxygen adsorption, a fresh cleaned copper surface would quickly oxidize at room temperature and develops a 1…2.5 nm thick film after a few seconds [1]. Molecular oxygen will be dissociated and bonded to free copper atoms at the metallic surface. The collision rate between the a specific surface area and molecular oxygen can be calculated according to [4]:

$$\text{collision rate} = \frac{p}{\sqrt{2\pi \cdot m \cdot k_B \cdot T}} \tag{51}$$

The variables are pressure p, gas molecule mass m, Boltzmann constant k_B and absolute temperature T. For a temperature of 20 °C a collision rate of $\approx 2.72 \cdot 10^{23}$ oxygen molecules / $mm^2 \cdot s$ can be determined according to Eq. (51) in common air. Although only 10 % of oxygen molecules would be directly bonded on the copper surface. A monolayer appears after a few seconds [1]. A closed film appears after a few minutes and the oxide growth rate will be reduced [26], [34], [35]. For a period of days or weeks after, growth ratios found in the literature are partially not consistent. However, all research confirms decreasing growth ratios at a constant temperature.

For the considered inert gas environment (Fig. 31) with Eq. (51) and 20 °C ambient temperature a collision rate of $\approx 2.18 \cdot 10^{19} - 8.16 \cdot 10^{20}$ directly bonded oxygen molecules / $mm^2 \cdot s$ would result. The unit cell of copper(I)oxide Cu_2O is about 0.42696 nm [29]. For an assumed area of 1 mm^2, $\approx 5.49 \cdot 10^{12}$ unit cells would be required for a monolayer. It can be seen from the determined collision rate that enough oxygen would be present to develop a monolayer. The resulting film resistance would be very low and only thin film tunneling can occur.

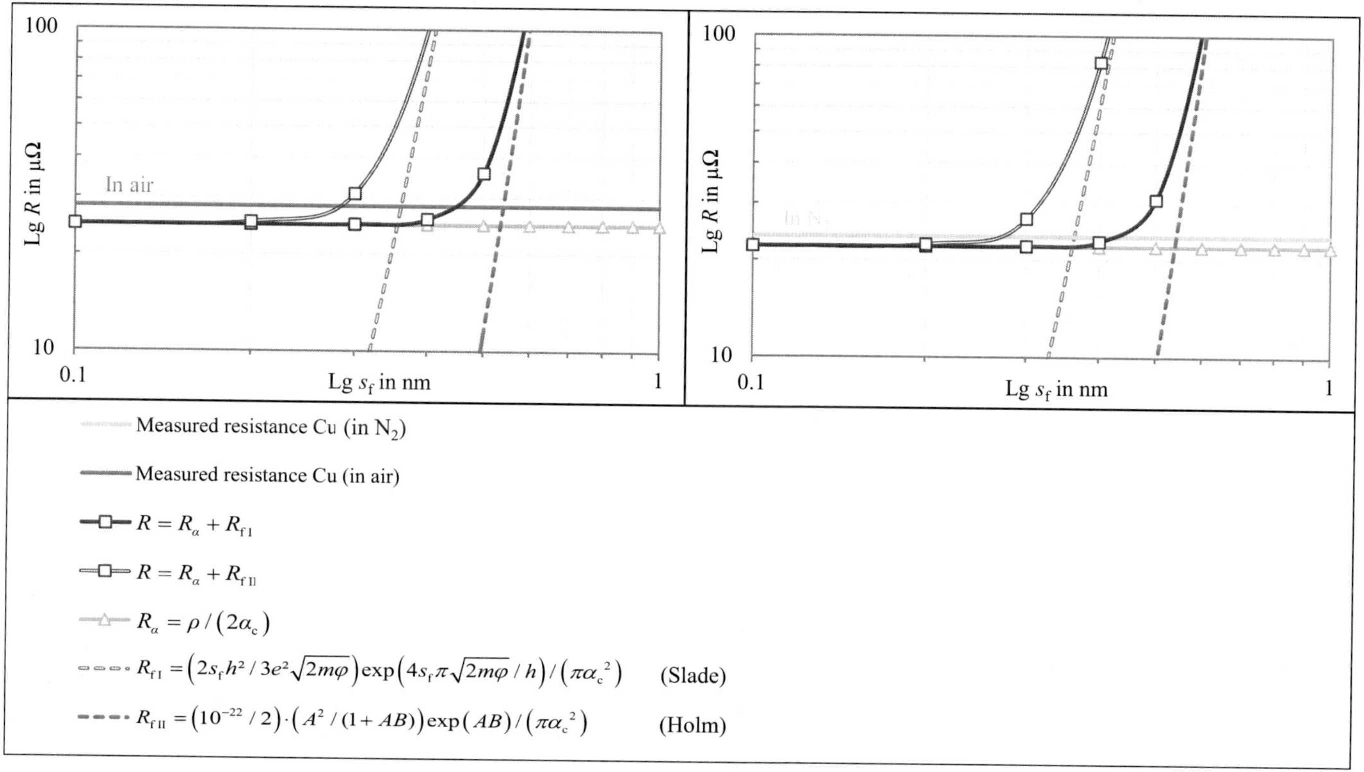

Fig. 89: Comparison of measured and calculated contact resistance of crossed copper rods, resistances determined by Eqs. (12) – (15) with the data from Tab. 8, Fig. 27 and [130]

The results in Fig. 89 show, that a small film resistance can be theoretical possible in air and even in N_2. The approach of Holm [1] results in slightly thicker films than the approach of Slade [3]. Depending on the selected approach, a possible film might be in the range of ≈ 0.25 … 0.45 nm. The differences between the measured resistance of crossed coppers rods in air and in N_2 are not significant (Fig. 90). Nearly all measured contact resistances are close or very close to the theoretical boundary. Moreover, the results demonstrate the lower influence of initial oxide film formation on copper surfaces for the contact resistance. The measurement principle and methodology was successful verified in N_2 inert gas atmosphere.

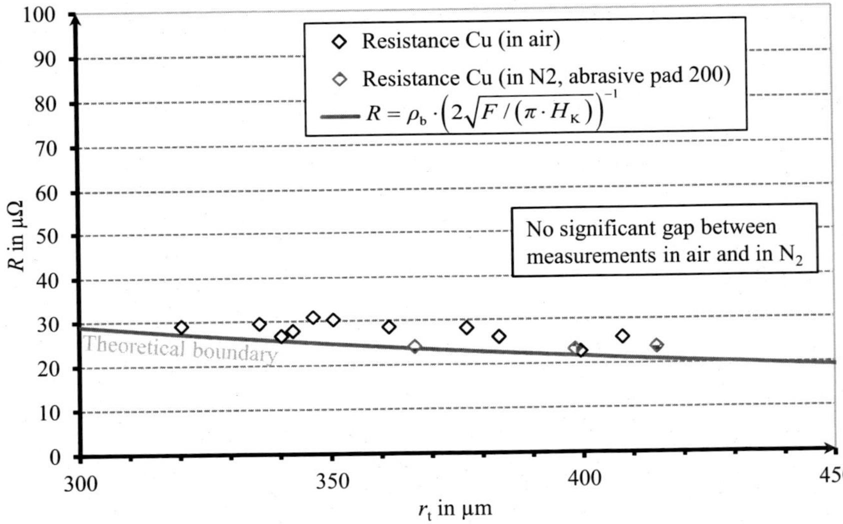

Fig. 90: Contact resistance R dependent on the radius of the plastic deformed area r_t plotted for individual measurements of copper as well as R determined by Eqs. (24) and (33) with the data of Cu from [78] and Tab. 8 and variable force F

Appendix G

Surface Oxidation

In Fig. 91, several areas emerge from the microscopic analysis. The distinction between the areas was done by color and the mechanical imprint of the other contact partner. Areas with a black, thick film (A) were obviously. The imprint of the other contact partner was observed at the plastic deformed area with thick oxide film (C). Parts of the thick oxide films were removed, possibly by mechanical forces during the opening process. In radial direction towards the contact area center, the black thick oxide films is merging into a red-brownish oxide film (D). Depending on the distance to the contact area center, eventually this film might become thinner.

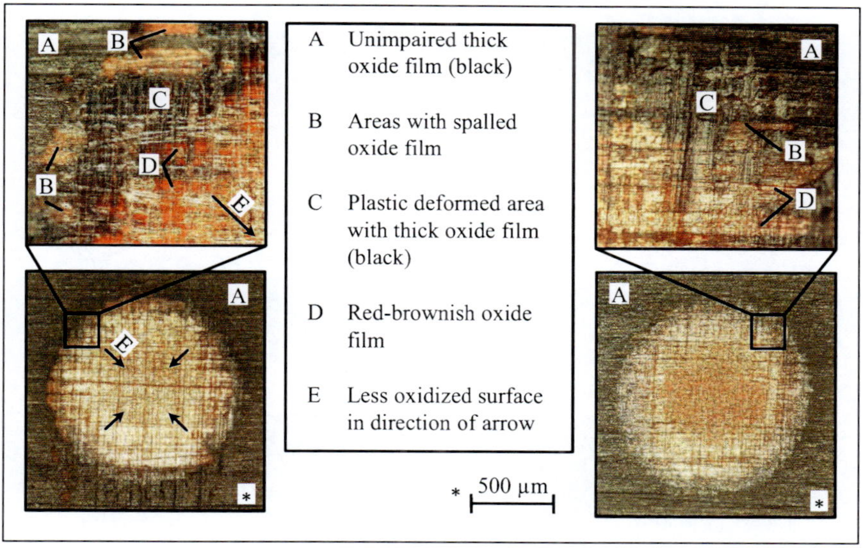

Fig. 91: Microscopic images of the contact area of crossed copper rods after 1000 h heat treatment at 200 °C in oven

Appendix H

Long-Term Study at 90 °C

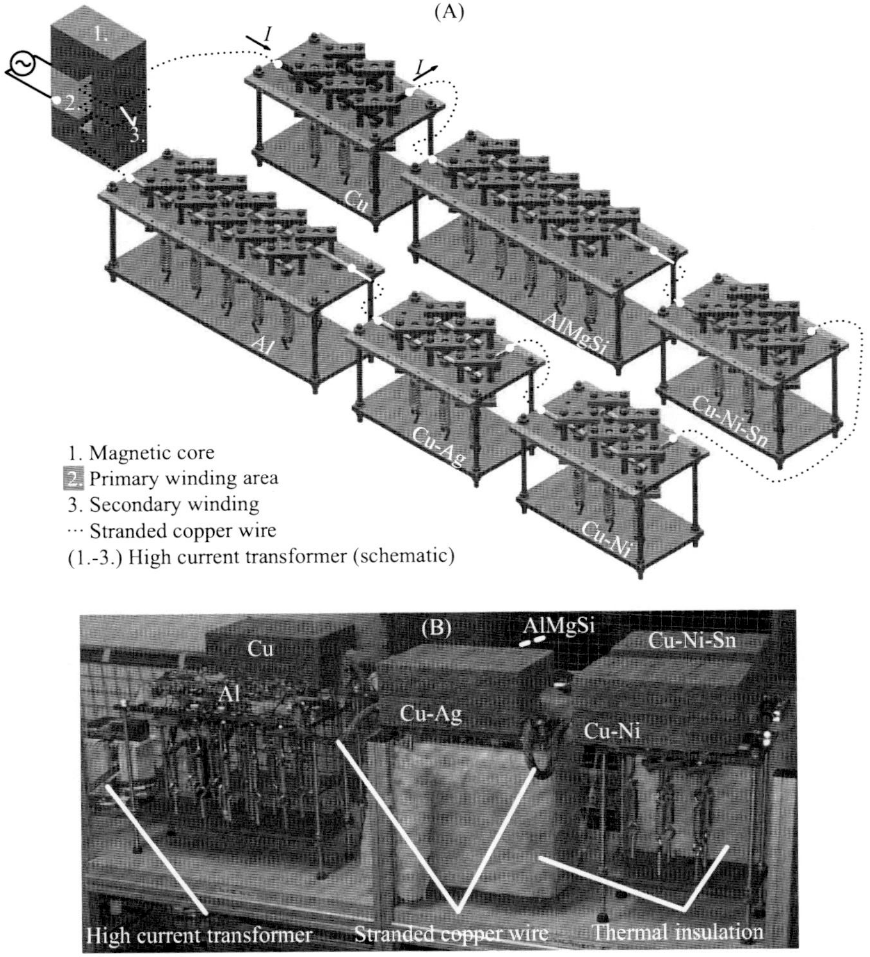

Fig. 92: Experimental setup long-term study in laboratory at 90 °C; schematic (A) and image (B); high current transformer RUE1400 So. UI132/72 Armco 1.4kVA, AC 230 V / AC 2.5V, 550 A, 50/60 Hz; variable transformer (not pictured): RRSP137 AC 230 V / AC 0-230 V, 10 A

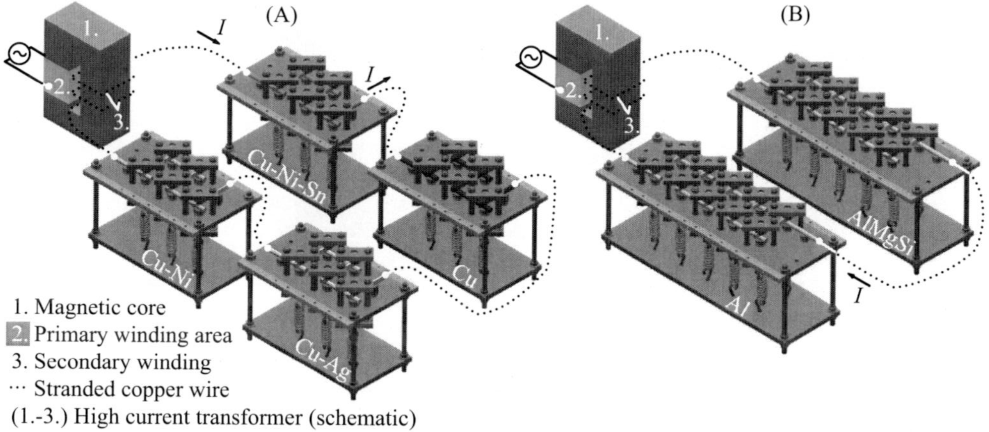

Fig. 93: Experimental setup long-term study in outdoor and tropical rainforest environment at 90 °C; high current transformer RUE1400 So. UI132/72 Armco 1.4kVA, AC 230 V / AC 2 V (A) or 2.5 V (B), 700 A (A) or 550 A (B), 50/60 Hz; variable transformer (not pictured): RRSP137 AC 230 V / AC 0-230 V, 10 A

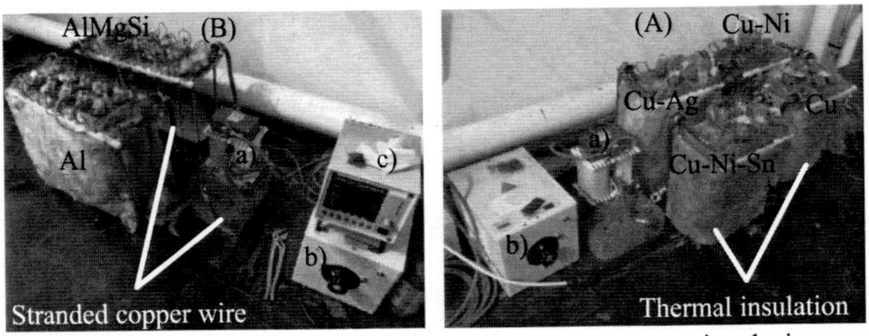

a) High current transformer
b) Variable transformer
c) Temperature measuring device

Fig. 94: Experimental setup long-term study in tropical rainforest environment at 90 °C, designation (A) and (B) according to Fig. 93

1. High current transformer (inside building)

Fig. 95: Experimental setup long-term study in outdoor environment at 90 °C, the power supply for the test assemblies within the shelter is inside the building and not visible on the image, designation (A) and (B) according to Fig. 93

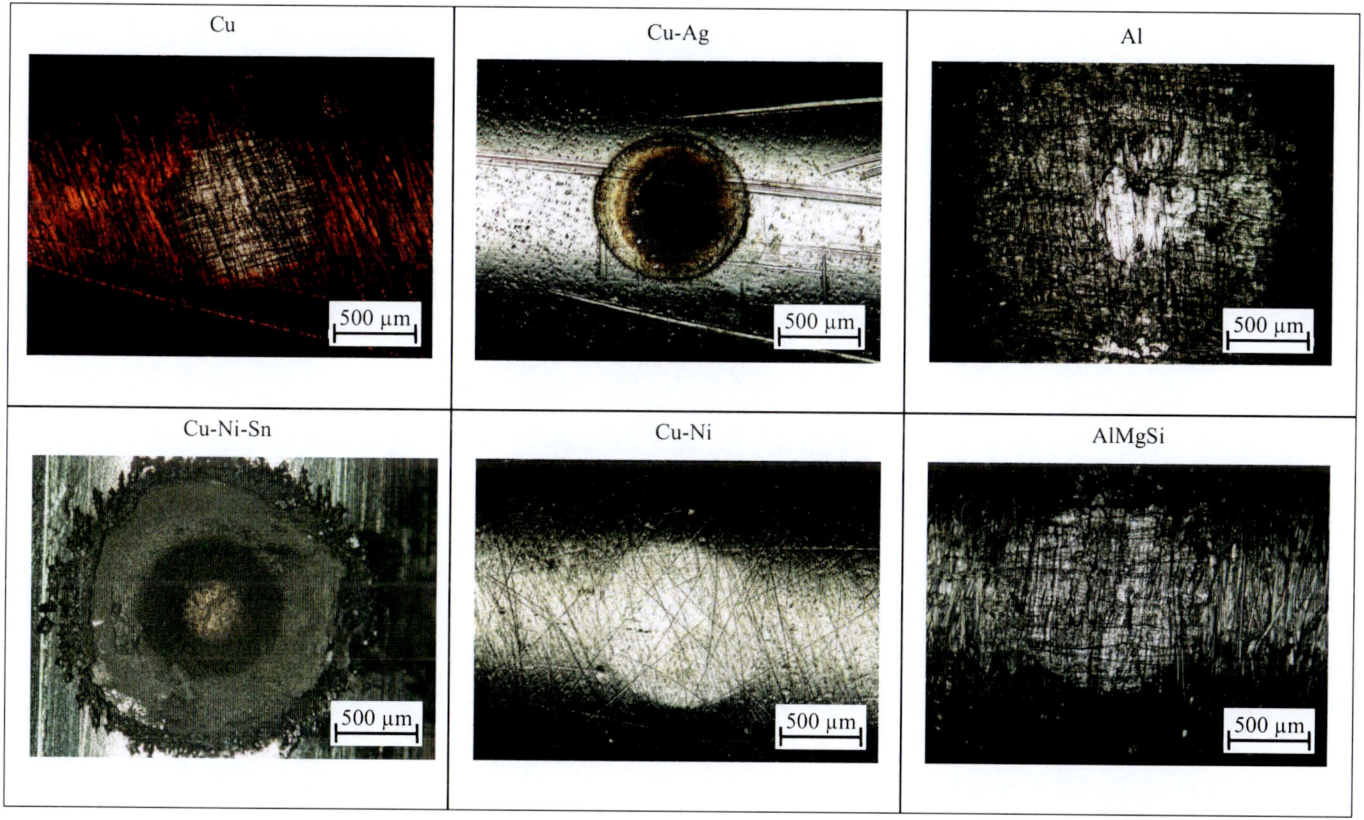

Fig. 96: Microscopic image of plastic deformed radii $r_t(t = 12000\ h)$ of laboratory environment at 90 °C

Long-Term Study 200 °C

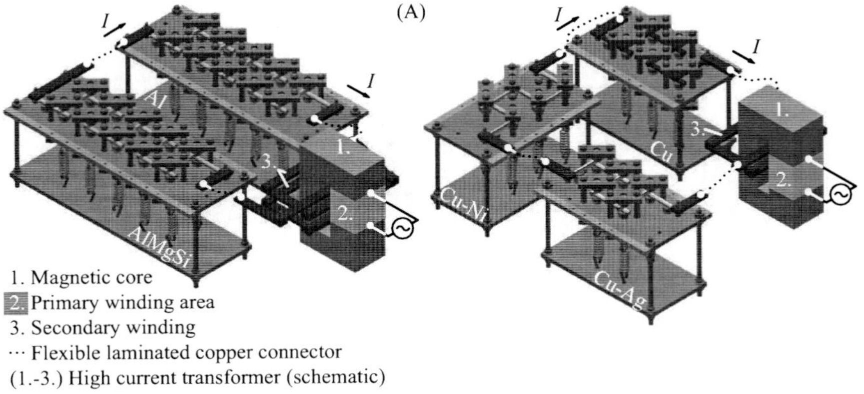

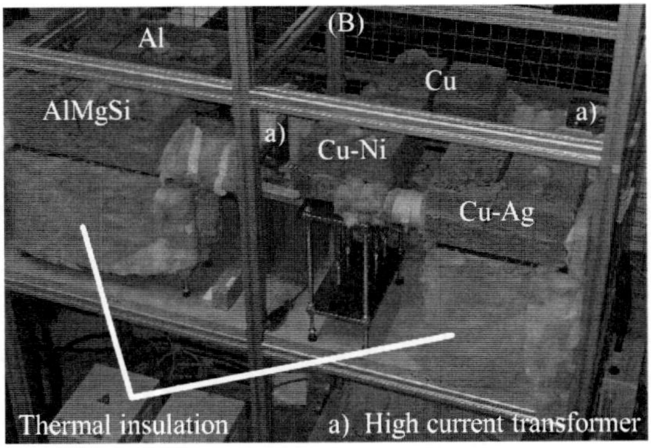

Fig. 97: Experimental setup long-term study in laboratory at 200 °C schematic (A) and image (B); high current transformer RUE1400 So. UI132/72 Armco 1.4 kVA, AC 230 V / AC 2 V or 2.5 V, 700 A or 550 A, 50/60 Hz; variable transformer (not pictured): RRSP137 AC 230 V / AC 0-230 V, 10 A

Tab. 26: Test circuits with applied current *I*

Test circuit	Fig. 97 Cu, Cu-Ag, Cu-Ni	Fig. 97 Al, AlMgSi	Fig. 92
ϑ in °C	*I* in A		
200	395 ± 30	521 ± 30	-
90	-	-	280 ± 20

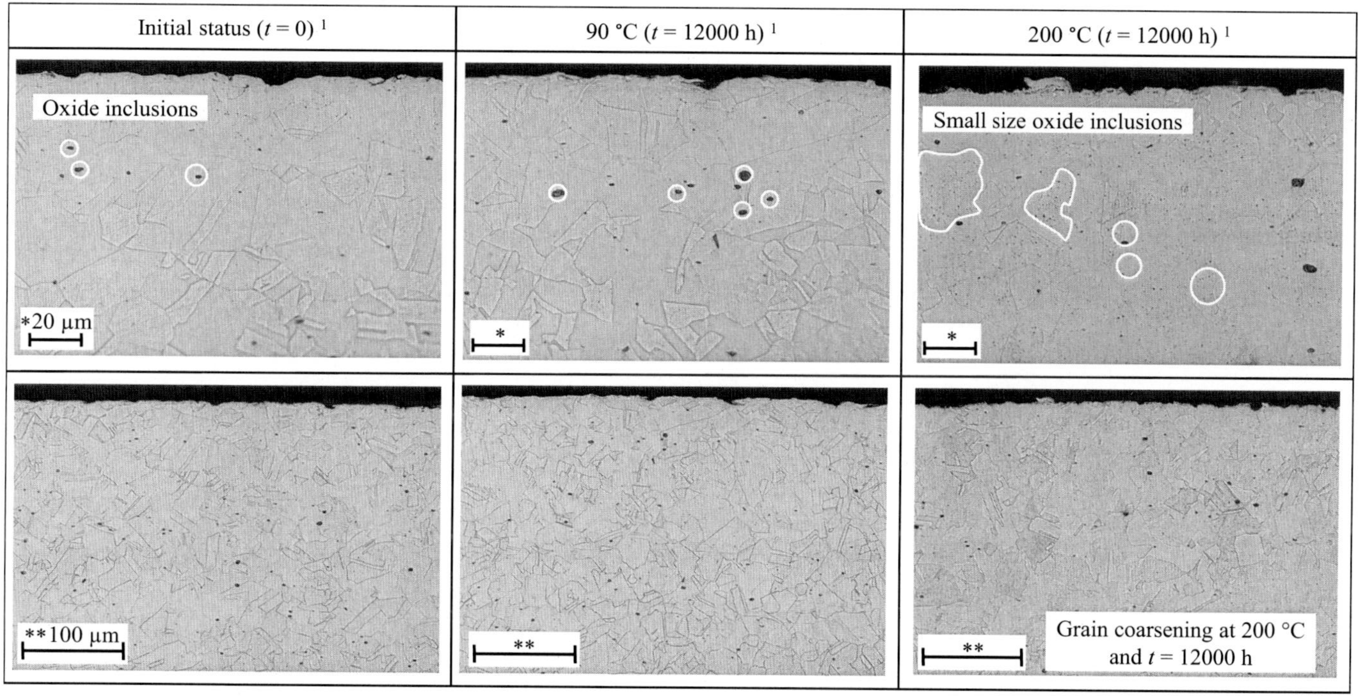

Fig. 98: Microstructure of Cu within the contact area dependent on time and temperature, [1] measurements done at IfWW, TUD

Tab. 27: Measured hardness dependent on material, temperature and operating time at the cross section area

Hardness H^* (measurements done at IfWW, TUD)											
	Initial state, $t = 0$			Laboratory 90 °C, $t = 1.4$ a			Laboratory 200 °C, $t = 1.4$ a			Measurement…	
	Mean	Max	Min	Mean	Max	Min	Mean	Max	Min	… method [1]	… position [2]
Cu	109	111	106	112.0	113.0	111.0	73	80	70	HV 10	Material
Cu-Ag [3]	127	128	125	125	128	122	126	128	122	HV0.001	Coating, load point
Cu-Ag [3]	112	113	108	80	82	77	78	84	76	HV0.001	Coating, reference point
Cu-Ni	484	514	464	552	572	542	223	228	220	HV0.001	Coating, load point
Cu-Ni	586	606	572	560	572	542	256	264	254	HV0.001	Coating, reference point
Cu-Ni-Sn	12	12	11	-	-	-	-	-	-	HV0.001	Coating, load point
Cu-Ni-Sn	13	13	13	13	13	12	-	-	-	HV0.001	Coating, reference point
Al	44	47	41	43	48	41	22	23	22	HV 10	Material
AlMgSi [4]	75	77	72	89	91	86	31	31	31	HV 10	Material

[1] The hardness measurement proceeding is based on the referenced method which is not equal to measurements according to standards. Especially the measured hardness of plated surface consist of a mixed hardness by substrate, coating and/or sublayer. In this study, the here measured hardness of plated surfaces can be only used for comparison between the same material systems.

[2] The measurement position of the coating was done within the contact area (load point) and at a reference point apart from the contact area.

[3] The hardness of silver declines due to the possible influence of the thermal induced recrystallization processes. This can be seen when comparing the hardness of the reference point at initial and aged state. At the load point the hardness is similar at all states. The recrystallization results in grain coarsening and occur as long as a thermal load is applied. Subsequently after powering down the test circuits and reaching ambient temperature, the still applied mechanical load will possibly transform the new developed grains. The outcome might be a grain structure similar to that at $t = 0$ and therefore similar hardness.

[4] During fabrication, AlMgSi (EN AW-6060 T6) undergo a heat treating process to develop better mechanical properties such as higher strength. This process is typically not performed until the maximum hardness is reached. Therefore, the aluminum hardness $H^*(t = 0)$ increased to $H^*(t = 1.4\ \text{a}, \vartheta = 90\ °\text{C})$ during thermal operation. The heat treating process is dependent on time, temperature and material. At $\vartheta = 200$ °C and $t = 1.4$ a, the material has exceeded the hardness maximum and reached an aged state. The hardness decreased to $H^*(t = 1.4\ \text{a}, \vartheta = 200\ °\text{C})$.

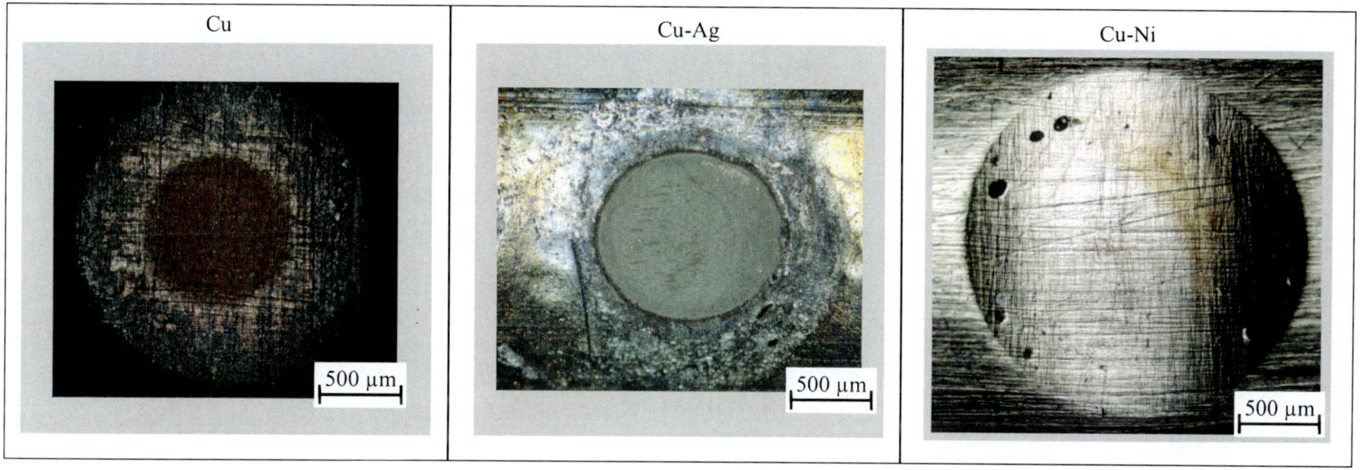

Fig. 99: Microscopic image of plastic deformed radii $r_{t}(t = 12000\ \text{h})$ for Cu, Cu-Ag and Cu-Ni in laboratory environment at 200 °C

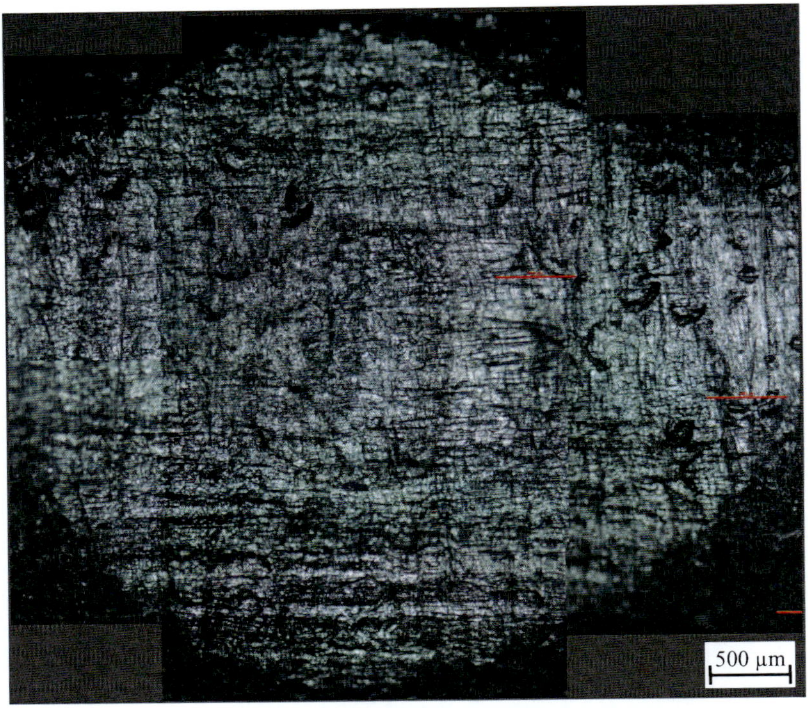

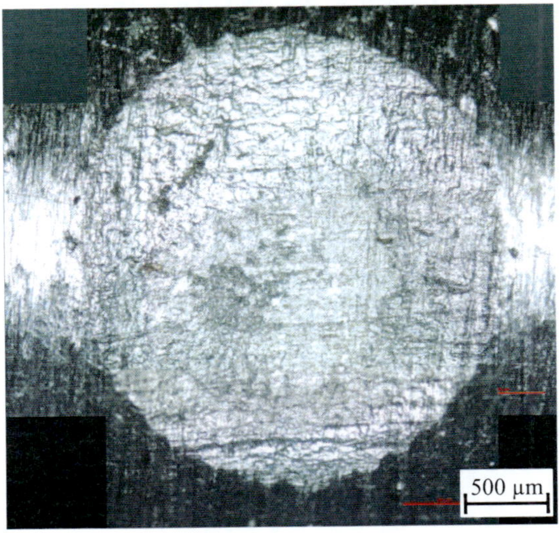

Fig. 100: Microscopic images of plastic deformed radii $r_t(t = 12000$ h) Al (superior) and AlMgSi (inferior) in laboratory environment at 200 °C

Long-Term Study at Low Forces in Oven with 200 °C

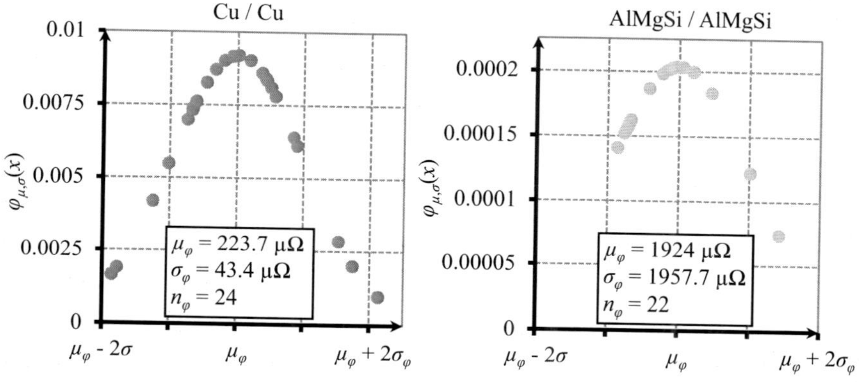

Fig. 101: Density function $\varphi_{\mu,\sigma}(x)$ with normal distribution of measured contact resistance of crossed rods (experimental setup illustrated in Fig. 49)

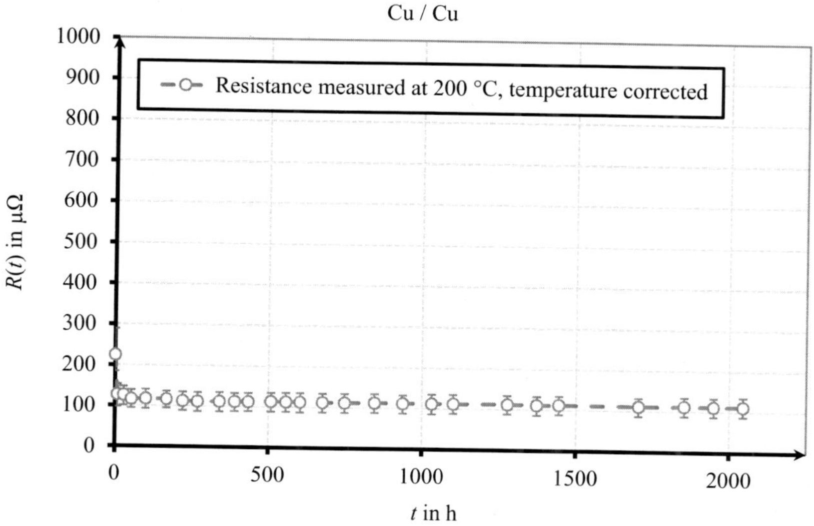

Fig. 102: Contact resistance $R(t)$ of crossed copper rods measured at 200 °C dependent on time and temperature, temperature correction according to Eq. (30)

Appendix J

Contact Resistance and Axial Film Growth

The initial resistance $R(t=0)$ of an individual a-spot with bulk material ρ_b and spot radius a where no films are present is as follows:

$$R(t=0)=\frac{\rho_b}{2a} \tag{52}$$

An additional film resistance with film thickness s_f and resistivity ρ_f has to be added at aged conditions:

$$R(t)=\frac{\rho_b}{2a}+\frac{\rho_f\cdot 2s_f}{\pi\cdot a^2} \tag{53}$$

The ratio of the relative degradation of the contact resistance $R(t)\,/\,R(t=0)$ is assumed to be between 2…5. Eq. (53) is now transformed to obtain the film thickness s_f dependent on the axial film growth:

$$s_f=\left(\frac{R(t)}{R(t=0)}-1\right)\cdot\frac{\rho_b\cdot\pi\cdot a}{4\rho_f} \tag{54}$$

The spot radius a can be replaced by the radius of the plastic deformed area r_t according to the findings in Chapter 6.3 and Appendix D. The principle approach was adjusted to each material combination (Tab. 29).

Tab. 28: Used bulk and film resistivity dependent on temperature as well as initial radii

			20 °C	200 °C		20 °C		200 °C
$r_t(t=0)$ in µm		ρ_b in Ωm			ρ_f in Ωm			
Cu / Cu	351	Cu	1.7E-08	3.0E-08	Cu_2O	5.5E+04	Cu_2O	6.0E+02
Al / Al	527	Al	2.8E-08	4.8E-08	Al_2O_3	9.7E+13	Al_2O_3	3.0E+12
AlMgSi / AlMgSi	428	AlMgSi	3.2E-08	5.4E-08		9.0E+13		3.0E+12
Cu-Ag / Ag-Cu	374	Ag	1.7E-08	2.9E-08	α-Ag_2S	3.9E+00	β-Ag_2S	3.2E-05
Cu-Ni / Ni-Cu	367	Ni	6.8E-08	1.5E-07	NiO	1.0E+11	NiO	6.1E+07
Cu-Ni-Sn / Sn-Ni-Cu	548	Sn	1.2E-07	2.1E-07	SnO	1.0E+02	SnO_2	5.0E-04

Tab. 29: Equations for film thickness s_f and ratio $R(t) / R(t = 0)$ for respective material combination

	Cu, Al, AlMgSi	Cu-Ag	Cu-Ni, Cu-Ni-Sn
$R(t=0) =$	$\frac{\rho_b}{2r_t}$ (55)	$\frac{\rho_b}{2r_t} + \frac{1}{\pi \cdot r_t} \cdot (\rho_b - \rho_{coa}) \cdot \arctan \frac{s_{coa}}{r_t}$ (59)	$\frac{\rho_b}{2r_t} + 2 \cdot \frac{\rho_{coa1} \cdot s_1}{\pi \cdot r_t^2} + 2 \cdot \frac{\rho_{coa2} \cdot s_2}{\pi \cdot r_t^2}$ (63)
$R(t) =$	$\frac{\rho_b}{2r_t} + \frac{\rho_f \cdot 2s_f}{\pi \cdot r_t^2}$ (56)	$\frac{\rho_b}{2r_t} + \frac{1}{\pi \cdot r_t} \cdot (\rho_b - \rho_{coa}) \cdot \arctan \frac{s_{coa}}{r_t} + \frac{\rho_f \cdot 2s_f}{\pi \cdot r_t^2}$ (60)	$\frac{\rho_b}{2r_t} + 2 \cdot \frac{\rho_{coa1} \cdot s_1}{\pi \cdot r_t^2} + 2 \cdot \frac{\rho_{coa2} \cdot s_2}{\pi \cdot r_t^2} + \frac{\rho_f \cdot 2s_f}{\pi \cdot r_t^2}$ (64)
$\frac{R(t)}{R(t=0)} =$	$1 + \frac{4\rho_f \cdot s_f}{\rho_b \cdot \pi \cdot r_t}$ (57)	$1 + \frac{\rho_f \cdot 2s_f}{r_t \cdot \left(\frac{\pi \cdot \rho_b}{2} + (\rho_b - \rho_{coa}) \cdot \arctan \frac{s_{coa}}{r_t} \right)}$ (61)	$1 + \frac{\rho_f \cdot s_f}{\frac{\pi \cdot r_t \cdot \rho_b}{4} + \rho_{coa1} \cdot s_1 + \rho_{coa2} \cdot s_2}$ (65)
$s_f =$	$\left(\frac{R(t)}{R(t=0)} - 1 \right) \cdot \frac{\rho_b \cdot \pi \cdot r_t}{4\rho_f}$ (58)	$\left(\frac{R(t)}{R(t=0)} - 1 \right) \cdot \frac{r_t}{2\rho_f} \cdot \left(\frac{\pi \cdot \rho_b}{2} + (\rho_b - \rho_{coa}) \cdot \arctan \frac{s_{coa}}{r_t} \right)$ (62)	$\left(\frac{R(t)}{R(t=0)} - 1 \right) \cdot \frac{1}{\rho_f} \cdot \left(\frac{\pi \cdot r_t \cdot \rho_b}{4} + \rho_{coa1} \cdot s_1 + \rho_{coa2} \cdot s_2 \right)$ (66)

Tab. 30: Theoretical thickness of axial developed films dependent on the degradation $R(t) / R(t = 0)$ according to Eqs. (55)-(66), temperature reference $\vartheta = 200$ °C

		Cu / Cu	Cu-Ag / Ag-Cu	Cu-Ni / Ni-Cu	Cu-Ni-Sn / Sn-Ni-Cu	Al / Al	AlMgSi / AlMgSi
		s_f in nm					
$\frac{R(t)}{R(t=0)}$	2	1.4E-07	2.7E+00	1.8E-13	2.6E-01	6.7E-17	3.5E-17
	5	5.5E-07	1.1E+01	7.4E-13	1.1E+00	2.7E-16	1.4E-16

Contact Resistance and Radial Film Growth

The film resistance $R_d \| R_f'$ can be defined as follows:

$$R_{\text{d}} \| R_{\text{f}}' = \frac{\dfrac{\sigma_{\text{f}}}{\pi \cdot (r^2 - a^2)} \cdot \dfrac{\rho_{\text{b}} \cdot 2s_{\text{d}}}{\pi \cdot a^2}}{\dfrac{\sigma_{\text{f}}}{\pi \cdot (r^2 - a^2)} + \dfrac{\rho_{\text{b}} \cdot 2s_{\text{d}}}{\pi \cdot a^2}} \tag{67}$$

After transformation Eq. (68) results:

$$R_{\text{d}} \| R_{\text{f}}' = \frac{1}{\pi \cdot \left(\dfrac{a^2}{\rho_{\text{b}} \cdot 2s_{\text{d}}} + \dfrac{r^2 - a^2}{\sigma_{\text{f}}} \right)} \tag{68}$$

Impact of Anisotropic Film Growth on the Current Distribution

The integration of Eq. (8) was performed with polar coordinates and $\mathrm{d}A = r\,\mathrm{d}r\,\mathrm{d}\varphi$.

$$I(r_{xy}) = \frac{I}{2\pi \cdot a} \int_0^{2\pi} \int_0^{r_{xy}} \frac{r}{\sqrt{a^2 - r^2}}\,\mathrm{d}r\,\mathrm{d}\varphi \tag{39}$$

To solve Eq. (39), integration by parts with Eq. (70) was used:

$$\int uv'\mathrm{d}x = uv - \int u'v\,\mathrm{d}x \qquad u = r \qquad v' = \frac{1}{\sqrt{a^2 - r^2}} \tag{69}$$

$$\int uv'\mathrm{d}x = r \cdot \arcsin\frac{r}{a} - \int \arcsin\frac{r}{a}\,\mathrm{d}r \tag{70}$$

$$\int uv'\mathrm{d}x = r \cdot \arcsin\frac{r}{a} - \left(r \cdot \arcsin\frac{r}{a} + \sqrt{a^2 - r^2} \right) + C = -\sqrt{a^2 - r^2} + C \tag{71}$$

$$I(r_{xy}) = \frac{I}{2\pi \cdot a} \int_0^{2\pi} \left[-\sqrt{a^2 - r^2} \right]_0^{r_{xy}} \mathrm{d}\varphi = \frac{I}{2\pi \cdot a} \int_0^{2\pi} \left(-\sqrt{a^2 - r_{xy}^{\ 2}} + a \right) \mathrm{d}\varphi \tag{72}$$

$$I(r_{xy}) = \frac{I}{2\pi \cdot a} \cdot \left[\left(-\sqrt{a^2 - r_{xy}^{\ 2}} + a \right) \cdot \varphi \right]_0^{2\pi} = I - \frac{I}{a} \cdot \sqrt{a^2 - r_{xy}^{\ 2}} \tag{73}$$

Tab. 31: Approaches for area A_i, mean current I_i and mean current density J_i dependent on radius r_{xyi}

	r_{xyi}	Area A_i	Current I_i Eq. (40)
I	$0 \ldots 0.5 \cdot a$	$0.5^2 \cdot \pi \cdot a^2$	$I(r_{xy} = 0.5 \cdot a)$ - $I(r_{xy} = 0)$
II	$0.5 \cdot a \ldots 0.75 \cdot a$	$(0.75^2 - 0.5^2) \cdot \pi \cdot a^2$	$I(r_{xy} = 0.75 \cdot a)$ - $I(r_{xy} = 0.5 \cdot a)$
III	$0.75 \cdot a \ldots a$	$(1 - 0.75^2) \cdot \pi \cdot a^2$	$I(r_{xy} = a)$ - $I(r_{xy} = 0.75 \cdot a)$

Tab. 32: Area A_i, mean current I_i, Eq. (40), and mean current density J_i dependent on radius r_{xyi}

	Zonal area I ($r_{xyI} = 0\ldots0.5\cdot a'$)			Zonal area II ($r_{xyII} = 0.5\cdot a'\ldots0.75\cdot a'$)			Zonal area III ($r_{xyIII} = 0.75\cdot a'\ldots a'$)		
a'	A_I	I_I	$J_I = I_I / A_I$	A_{II}	I_{II} (40)	$J_{II} = I_I / A_I$	A_{III}	I_{III} (40)	$J_{III} = I_I / A_I$
a	$0.25\cdot\pi\, a^2$	$0.13\cdot I$	$0.54\cdot I / (\pi\cdot a^2)$	$0.31\cdot\pi\, a^2$	$0.2\cdot I$	$0.65\cdot I / (\pi\cdot a^2)$	$0.44\cdot\pi\, a^2$	$0.66\cdot I$	$1.51\cdot I / (\pi\cdot a^2)$
$0.9\cdot a$	$0.2\cdot\pi\, a^2$	$0.13\cdot I$	$0.66\cdot I / (\pi\cdot a^2)$	$0.25\cdot\pi\, a^2$	$0.2\cdot I$	$0.81\cdot I / (\pi\cdot a^2)$	$0.35\cdot\pi\, a^2$	$0.66\cdot I$	$1.87\cdot I / (\pi\cdot a^2)$
$0.8\cdot a$	$0.16\cdot\pi\, a^2$	$0.13\cdot I$	$0.84\cdot I / (\pi\cdot a^2)$	$0.2\cdot\pi\, a^2$	$0.2\cdot I$	$1.02\cdot I / (\pi\cdot a^2)$	$0.28\cdot\pi\, a^2$	$0.66\cdot I$	$2.3\cdot I / (\pi\cdot a^2)$
$0.7\cdot a$	$0.12\cdot\pi\, a^2$	$0.13\cdot I$	$1.09\cdot I / (\pi\cdot a^2)$	$0.15\cdot\pi\, a^2$	$0.2\cdot I$	$1.34\cdot I / (\pi\cdot a^2)$	$0.21\cdot\pi\, a^2$	$0.66\cdot I$	$3.09\cdot I / (\pi\cdot a^2)$
$0.6\cdot a$	$0.09\cdot\pi\, a^2$	$0.13\cdot I$	$1.49\cdot I / (\pi\cdot a^2)$	$0.11\cdot\pi\, a^2$	$0.2\cdot I$	$1.82\cdot I / (\pi\cdot a^2)$	$0.16\cdot\pi\, a^2$	$0.66\cdot I$	$4.2\cdot I / (\pi\cdot a^2)$
$0.5\cdot a$	$0.06\cdot\pi\, a^2$	$0.13\cdot I$	$2.14\cdot I / (\pi\cdot a^2)$	$0.08\cdot\pi\, a^2$	$0.2\cdot I$	$2.62\cdot I / (\pi\cdot a^2)$	$0.11\cdot\pi\, a^2$	$0.66\cdot I$	$6.05\cdot I / (\pi\cdot a^2)$
$0.4\cdot a$	$0.04\cdot\pi\, a^2$	$0.13\cdot I$	$3.35\cdot I / (\pi\cdot a^2)$	$0.05\cdot\pi\, a^2$	$0.2\cdot I$	$4.09\cdot I / (\pi\cdot a^2)$	$0.07\cdot\pi\, a^2$	$0.66\cdot I$	$9.45\cdot I / (\pi\cdot a^2)$
$0.3\cdot a$	$0.02\cdot\pi\, a^2$	$0.13\cdot I$	$5.95\cdot I / (\pi\cdot a^2)$	$0.03\cdot\pi\, a^2$	$0.2\cdot I$	$7.27\cdot I / (\pi\cdot a^2)$	$0.04\cdot\pi\, a^2$	$0.66\cdot I$	$16.8\cdot I / (\pi\cdot a^2)$
$0.2\cdot a$	$0.01\cdot\pi\, a^2$	$0.13\cdot I$	$13.4\cdot I / (\pi\cdot a^2)$	$0.01\cdot\pi\, a^2$	$0.2\cdot I$	$16.37\cdot I / (\pi\cdot a^2)$	$0.02\cdot\pi\, a^2$	$0.66\cdot I$	$37.8\cdot I / (\pi\cdot a^2)$
$0.1\cdot a$	$2.5E\text{-}03\cdot\pi\, a^2$	$0.13\cdot I$	$53.59\cdot I / (\pi\cdot a^2)$	$3.1E\text{-}03\cdot\pi\, a^2$	$0.2\cdot I$	$65.47\cdot I / (\pi\cdot a^2)$	$4.4E\text{-}03\cdot\pi\, a^2$	$0.66\cdot I$	$151.19\cdot I / (\pi\cdot a^2)$
$0.01\cdot a$	$2.5E\text{-}05\cdot\pi\, a^2$	$0.13\cdot I$	$5359\cdot I / (\pi\cdot a^2)$	$3.1E\text{-}05\cdot\pi\, a^2$	$0.2\cdot I$	$6547\cdot I / (\pi\cdot a^2)$	$4.4E\text{-}05\cdot\pi\, a^2$	$0.66\cdot I$	$15119\cdot I / (\pi\cdot a^2)$

Multiple Contact Spots

$$I_{ia} = \frac{I_i}{n_i} = \frac{I_i}{\frac{A_i}{A}\cdot n} \tag{74}$$

Tab. 33: Area A_i, mean current I_i, mean current per spot I_{ia} dependent on radius $r_{xyi,}$

	r_{xyi}	Area A_i	Mean Current I_i	Mean Current I_{ia} per spot, Eq. (74)
I	$0\ldots0.5\cdot\alpha_c$	$0.25\cdot\pi\cdot\alpha_c^2$	$I(r_{xy} = 0.5\,\alpha_c)$ - $I(r_{xy} = 0)$	$0.52\cdot I / n$
II	$0.5\cdot\alpha_c\ldots0.75\cdot\alpha_c$	$(0.75^2 - 0.5^2)\cdot\pi\cdot\alpha_c^2$	$I(r_{xy} = 0.75\cdot\alpha_c)$ - $I(r_{xy} = 0.5\cdot\alpha_c)$	$0.65\cdot I / n$
III	$0.75\cdot\alpha_c\ldots\alpha_c$	$(1 - 0.75^2)\cdot\pi\cdot\alpha_c^2$	$I(r_{xy} = \alpha_c)$ - $I(r_{xy} = 0.75\cdot\alpha_c)$	$1.5\cdot I / n$

Tab. 34: Area A_i, mean current I_i, Eq. (40), mean current per spot I_{ia}, Eq. (74), mean current density J_i, mean current density per spot J_{ia} dependent on radius r_{xyi}

	Zonal area I ($r_{xy\mathrm{I}} = 0 \ldots 0.5 \cdot \alpha_c'$)				
α_c'	A_I	$\mathrm{I_I}$	$I_{\mathrm{I}a}$	$J_\mathrm{I} = I_\mathrm{I} / A_\mathrm{I}$	$J_{Ia} = I_{Ia} / (\pi \cdot a^2)$
α_c	$0.25 \cdot \pi\ \alpha_c^2$	$0.13 \cdot I$	$0.54 \cdot I / n$	$0.54 \cdot I / (\pi \cdot \alpha_c^2)$	$0.54 \cdot \mathrm{I} / (n \cdot \pi \cdot a^2)$
$0.9 \cdot \alpha_c$	$0.2 \cdot \pi\ \alpha_c^2$	$0.13 \cdot I$	$0.66 \cdot I / n$	$0.66 \cdot I / (\pi \cdot \alpha_c^2)$	$0.66 \cdot \mathrm{I} / (n \cdot \pi \cdot a^2)$
$0.8 \cdot \alpha_c$	$0.16 \cdot \pi\ \alpha_c^2$	$0.13 \cdot I$	$0.84 \cdot I / n$	$0.84 \cdot I / (\pi \cdot \alpha_c^2)$	$0.84 \cdot \mathrm{I} / (n \cdot \pi \cdot a^2)$
$0.7 \cdot \alpha_c$	$0.12 \cdot \pi\ \alpha_c^2$	$0.13 \cdot I$	$1.09 \cdot I / n$	$1.09 \cdot I / (\pi \cdot \alpha_c^2)$	$1.09 \cdot \mathrm{I} / (n \cdot \pi \cdot a^2)$
$0.6 \cdot \alpha_c$	$0.09 \cdot \pi\ \alpha_c^2$	$0.13 \cdot I$	$1.49 \cdot I / n$	$1.49 \cdot I / (\pi \cdot \alpha_c^2)$	$1.49 \cdot \mathrm{I} / (n \cdot \pi \cdot a^2)$
$0.5 \cdot \alpha_c$	$0.06 \cdot \pi\ \alpha_c^2$	$0.13 \cdot I$	$2.14 \cdot I / n$	$2.14 \cdot I / (\pi \cdot \alpha_c^2)$	$2.14 \cdot \mathrm{I} / (n \cdot \pi \cdot a^2)$
$0.4 \cdot \alpha_c$	$0.04 \cdot \pi\ a^2$	$0.13 \cdot I$	$3.35 \cdot I / n$	$3.35 \cdot I / (\pi \cdot \alpha_c^2)$	$3.35 \cdot \mathrm{I} / (n \cdot \pi \cdot a^2)$
$0.3 \cdot \alpha_c$	$0.02 \cdot \pi\ \alpha_c^2$	$0.13 \cdot I$	$5.95 \cdot I / n$	$5.95 \cdot I / (\pi \cdot \alpha_c^2)$	$5.95 \cdot \mathrm{I} / (n \cdot \pi \cdot a^2)$
$0.2 \cdot \alpha_c$	$0.01 \cdot \pi\ \alpha_c^2$	$0.13 \cdot I$	$13.4 \cdot I / n$	$13.4 \cdot I / (\pi \cdot \alpha_c^2)$	$13.4 \cdot \mathrm{I} / (n \cdot \pi \cdot a^2)$
$0.1 \cdot \alpha_c$	$2.5\mathrm{E}{-03} \cdot \pi\ \alpha_c^2$	$0.13 \cdot I$	$53.59 \cdot I / n$	$53.59 \cdot I / (\pi \cdot \alpha_c^2)$	$53.59 \cdot I / (n \cdot \pi \cdot a^2)$
$0.01 \cdot \alpha_c$	$2.5\mathrm{E}{-05} \cdot \pi\ \alpha_c^2$	$0.13 \cdot I$	$5359 \cdot I / n$	$5359 \cdot I / (\pi \cdot \alpha_c^2)$	$5359 \cdot \mathrm{I} / (n \cdot \pi \cdot a^2)$
	Zonal area II ($r_{xy\mathrm{II}} = 0.5 \cdot \alpha_c' \ldots 0.75 \cdot \alpha_c'$)				
α_c'	A_II	I_II	$I_{\mathrm{II}a}$	$J_\mathrm{II} = I_\mathrm{I} / A_\mathrm{I}$	$J_{IIa} = I_{IIa} / (\pi \cdot a^2)$
α_c	$0.31 \cdot \pi\ \alpha_c^2$	$0.2 \cdot I$	$0.65 \cdot I / n$	$0.65 \cdot I / (\pi \cdot \alpha_c^2)$	$0.65 \cdot \mathrm{I} / (n \cdot \pi \cdot a^2)$
$0.9 \cdot \alpha_c$	$0.25 \cdot \pi\ \alpha_c^2$	$0.2 \cdot I$	$0.81 \cdot I / n$	$0.81 \cdot I / (\pi \cdot \alpha_c^2)$	$0.81 \cdot \mathrm{I} / (n \cdot \pi \cdot a^2)$
$0.8 \cdot \alpha_c$	$0.2 \cdot \pi\ \alpha_c^2$	$0.2 \cdot I$	$1.02 \cdot I / n$	$1.02 \cdot I / (\pi \cdot \alpha_c^2)$	$1.02 \cdot \mathrm{I} / (n \cdot \pi \cdot a^2)$
$0.7 \cdot \alpha_c$	$0.15 \cdot \pi\ \alpha_c^2$	$0.2 \cdot I$	$1.34 \cdot I / n$	$1.34 \cdot I / (\pi \cdot \alpha_c^2)$	$1.34 \cdot \mathrm{I} / (n \cdot \pi \cdot a^2)$
$0.6 \cdot \alpha_c$	$0.11 \cdot \pi\ \alpha_c^2$	$0.2 \cdot I$	$1.82 \cdot I / n$	$1.82 \cdot I / (\pi \cdot \alpha_c^2)$	$1.82 \cdot \mathrm{I} / (n \cdot \pi \cdot a^2)$
$0.5 \cdot \alpha_c$	$0.08 \cdot \pi\ \alpha_c^2$	$0.2 \cdot I$	$2.62 \cdot I / n$	$2.62 \cdot I / (\pi \cdot \alpha_c^2)$	$2.62 \cdot \mathrm{I} / (n \cdot \pi \cdot a^2)$
$0.4 \cdot \alpha_c$	$0.05 \cdot \pi\ \alpha_c^2$	$0.2 \cdot I$	$4.09 \cdot I / n$	$4.09 \cdot I / (\pi \cdot \alpha_c^2)$	$4.09 \cdot \mathrm{I} / (n \cdot \pi \cdot a^2)$
$0.3 \cdot \alpha_c$	$0.03 \cdot \pi\ \alpha_c^2$	$0.2 \cdot I$	$7.27 \cdot I / n$	$7.27 \cdot I / (\pi \cdot \alpha_c^2)$	$7.27 \cdot \mathrm{I} / (n \cdot \pi \cdot a^2)$
$0.2 \cdot \alpha_c$	$0.01 \cdot \pi\ \alpha_c^2$	$0.2 \cdot I$	$16.37 \cdot I / n$	$16.37 \cdot I / (\pi \cdot \alpha_c^2)$	$16.37 \cdot I / (n \cdot \pi \cdot a^2)$
$0.1 \cdot \alpha_c$	$3.1\mathrm{E}{-03} \cdot \pi\ \alpha_c^2$	$0.2 \cdot I$	$65.47 \cdot I / n$	$65.47 \cdot I / (\pi \cdot \alpha_c^2)$	$65.47 \cdot I / (n \cdot \pi \cdot a^2)$
$0.01 \cdot \alpha_c$	$3.1\mathrm{E}{-05} \cdot \pi\ \alpha_c^2$	$0.2 \cdot I$	$6547 \cdot I / n$	$6547 \cdot I / (\pi \cdot \alpha_c^2)$	$6547 \cdot \mathrm{I} / (n \cdot \pi \cdot a^2)$
	Zonal area III ($r_{xy\mathrm{III}} = 0.75 \cdot \alpha_c' \ldots \alpha_c'$)				
α_c'	A_III	I_III	$I_{\mathrm{III}a}$	$J_\mathrm{III} = I_\mathrm{I} / A_\mathrm{I}$	$\mathrm{J_{IIIa}} = \mathrm{I_{IIIa}} / (\pi \cdot a^2)$
α_c	$0.44 \cdot \pi\ \alpha_c^2$	$0.66 \cdot I$	$1.51 \cdot I / n$	$1.51 \cdot I / (\pi \cdot \alpha_c^2)$	$1.51 \cdot \mathrm{I} / (n \cdot \pi \cdot a^2)$
$0.9 \cdot \alpha_c$	$0.35 \cdot \pi\ \alpha_c^2$	$0.66 \cdot I$	$1.87 \cdot I / n$	$1.87 \cdot I / (\pi \cdot \alpha_c^2)$	$1.87 \cdot \mathrm{I} / (n \cdot \pi \cdot a^2)$
$0.8 \cdot \alpha_c$	$0.28 \cdot \pi\ \alpha_c^2$	$0.66 \cdot I$	$2.36 \cdot I / n$	$2.3 \cdot I / (\pi \cdot \alpha_c^2)$	$2.3 \cdot \mathrm{I} / (n \cdot \pi \cdot a^2)$
$0.7 \cdot \alpha_c$	$0.21 \cdot \pi\ \alpha_c^2$	$0.66 \cdot I$	$3.09 \cdot I / n$	$3.09 \cdot I / (\pi \cdot \alpha_c^2)$	$3.09 \cdot \mathrm{I} / (n \cdot \pi \cdot a^2)$
$0.6 \cdot \alpha_c$	$0.16 \cdot \pi\ \alpha_c^2$	$0.66 \cdot I$	$4.2 \cdot I / n$	$4.20 \cdot I / (\pi \cdot \alpha_c^2)$	$4.2 \cdot \mathrm{I} / (n \cdot \pi \cdot a^2)$
$0.5 \cdot \alpha_c$	$0.11 \cdot \pi\ \alpha_c^2$	$0.66 \cdot I$	$6.05 \cdot I / n$	$6.05 \cdot I / (\pi \cdot \alpha_c^2)$	$6.05 \cdot \mathrm{I} / (n \cdot \pi \cdot a^2)$
$0.4 \cdot \alpha_c$	$0.07 \cdot \pi\ \alpha_c^2$	$0.66 \cdot I$	$9.45 \cdot I / n$	$9.45 \cdot I / (\pi \cdot \alpha_c^2)$	$9.45 \cdot \mathrm{I} / (\mathrm{n} \cdot \pi \cdot \mathrm{a}^2)$
$0.3 \cdot \alpha_c$	$0.04 \cdot \pi\ \alpha_c^2$	$0.66 \cdot I$	$16.8 \cdot I / n$	$16.8 \cdot I / (\pi \cdot \alpha_c^2)$	$16.8 \cdot \mathrm{I} / (n \cdot \pi \cdot a^2)$
$0.2 \cdot \alpha_c$	$0.02 \cdot \pi\ \alpha_c^2$	$0.66 \cdot I$	$37.8 \cdot I / n$	$37.8 \cdot I / (\pi \cdot \alpha_c^2)$	$37.8 \cdot \mathrm{I} / (n \cdot \pi \cdot a^2)$
$0.1 \cdot \alpha_c$	$4.4\mathrm{E}{-03} \cdot \pi\ \alpha_c^2$	$0.66 \cdot I$	$151.19 \cdot I / n$	$151.19 \cdot I / (\pi \cdot \alpha_c^2)$	$151.19 \cdot I / (n \cdot \pi \cdot a^2)$
$0.01 \cdot \alpha_c$	$4.4\mathrm{E}{-05} \cdot \pi\ \alpha_c^2$	$0.66 \cdot I$	$15119 \cdot I / n$	$15119 \cdot I / (\pi \cdot \alpha_c^2)$	$15119 \cdot I / (n \cdot \pi \cdot a^2)$

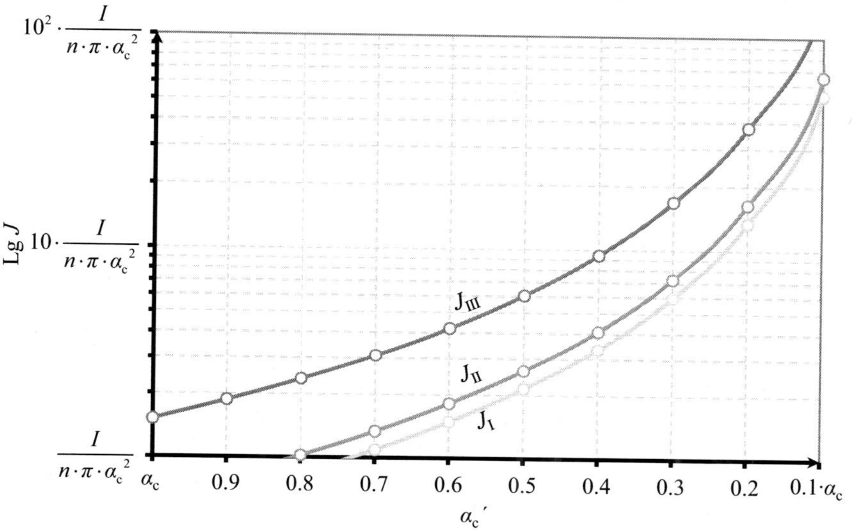

Fig. 103: Current density J depending on cluster radius α_c'

Appendix K

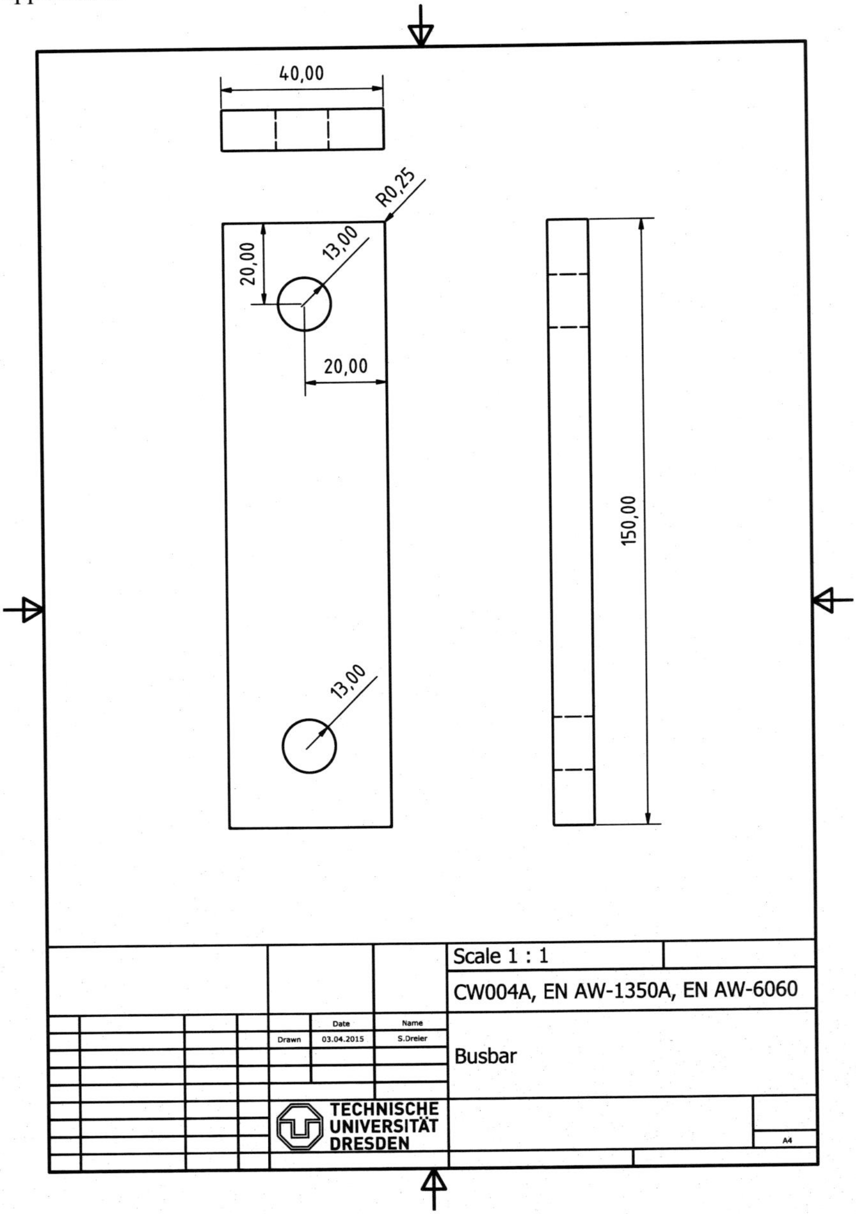

Fig. 104: Drawing busbar